Tropic

D0933381

Bernard A. Marcus
Genessee Community College
Batavia, NY

JONES AND BARTLETT PUBLISHERS
Sudbury, Massachusetts
BOSTON   TORONTO   LONDON   SINGAPORE

World Headquarters

Jones and Bartlett Publishers
40 Tall Pine Drive
Sudbury, MA 01776
978-443-5000
info@jbpub.com
www.jbpub.com

Jones and Bartlett Publishers
Canada
6339 Ormindale Way
Mississauga, Ontario L5V 1J2
Canada

Jones and Bartlett Publishers
International
Barb House, Barb Mews
London W6 7PA
United Kingdom

Jones and Bartlett's books and products are available through most bookstores and online booksellers. To contact Jones and Bartlett Publishers directly, call 800-832-0034, fax 978-443-8000, or visit our website www.jbpub.com.

Substantial discounts on bulk quantities of Jones and Bartlett's publications are available to corporations, professional associations, and other qualified organizations. For details and specific discount information, contact the special sales department at Jones and Bartlett via the above contact information or send an e-mail to specialsales@jbpub.com.

Production Credits:
Executive Editor, Science: Cathleen Sether
Managing Editor, Science: Dean W. DeChambeau
Acquisition Editor, Science: Shoshanna Goldberg
Associate Editor, Science: Molly Steinbach
Editorial Assistant, Science: Caroline Perry
Production Director: Amy Rose
Production Editor: Daniel Stone
V.P., Manufacturing and Inventory Control: Therese Connell
Senior Marketing Manager: Andrea DeFronzo
Cover Design: Kate Ternullo
Photo Research Manager: Kimberly L. Potvin
Cover Credit: © Alexander S. Heitcamp, ShutterStock, Inc.
Composition: Auburn Associates, Inc.
Printing and Binding: Malloy, Inc.
Cover Printing: Malloy, Inc.

Library of Congress Cataloging-in-Publication Data
Marcus, Bernard A.
  Tropical forests / Bernard A. Marcus.
    p. cm.
  Includes bibliographical references and index.
  ISBN-13: 978-0-7637-5434-1
  ISBN-10: 0-7637-5434-X
  1. Forests and forestry—Tropics. 2. Rainforest conservation. I. Title.
  SD247M37 2008
  577.34—dc22

                                                                        2008007280

6048
Printed in the United States of America
12 11 10 09 08 10 9 8 7 6 5 4 3 2 1

In memory of Herman S. Forest: Scholar, mentor, friend

Contents

Editor's Foreword

THIS VOLUME AND SERIES REPRESENT an endeavor by the author and editor to provide a start for those who wish to learn more about the ecosystems and biomes that support life on Earth. In addition to presenting information on these systems and the challenges they face, it offers references for further and more detailed studies to those who wish to avail themselves of such. Moreover, the author has done his best not only to make the material readable and understandable to those who do not have an extensive background in the environmental sciences, but also to illustrate the information with examples relative to most readers' experiences.

As series editor, it is my sincere hope that those using these books will find them not only informative, but enlightening and enjoyable as well. We have only one planet. It would be a shame to spoil it.

Bernard A. Marcus
Syracuse, NY

Preface

THE WORLD'S TROPICAL FORESTS ARE in danger. Not only the rainforests that capture many people's imaginations but also the dry forests that nobody in North America seems to have heard about.

The British scientist Sir Francis Bacon is credited with having said "Knowledge is power." The absence of knowledge about the tropical forests has contributed mightily toward their destruction.

My goal in writing this book was to share knowledge. In the past I did so by leading student trips to Central America. More than once I heard awed young adults marvel over how "awesome" the forests were and how "cool" it was that the indigenous people know how to extract so many useful products from the forests without destroying them. In addition, many were appalled by the forest destruction they saw. Regrettably, those days are behind me. But there is still forest to see and to save and species to discover and to preserve.

Fortunately, there are knowledgeable people already working toward those goals, and they can use help. My hope is that anyone who picks up this book will become sufficiently interested in the tropical forests to want to learn more about them. I believe that in most cases, the more one knows, the more one will want to help conserve and hopefully contribute to the restoration of these threatened ecosystems.

Bernard A. Marcus
Syracuse, NY

Introduction

BETWEEN THE TROPICS OF CANCER and Capricorn, roughly 24° north and south of the equator, the Earth is in constant summer. Seasonal changes in temperature don't occur much, and the winds blow warm, usually from the east. As they cross warm oceans, the winds evaporate moisture, which they carry toward the west. Eventually, the moisture-laden air blows across land, where it begins to rise. As it crosses more land, it rises higher and higher, often where it has to pass over hills and mountains. Where the land is forested, usually between 10° north and south of the equator, more moisture evaporates from trees and accumulates with that brought in from the oceans. As the humid air climbs higher and higher, it begins to encounter colder and colder air that causes it to cool. As it cools, its moisture begins to condense, and clouds form. Still, the clouds continue to climb until the air around them becomes too cold to hold the moisture any longer; at that point it starts to rain. Sometimes the rain lasts for a few minutes; sometimes it lasts for hours or days. In some parts of the world, it rains nearly every day, and the forests grow lush and green. Rainfall often exceeds 100 inches (250 cm) a year in places.

This is the land of the tropical rainforest, where trees grow tall and year round. Some trees may grow to heights in excess of 150 feet (45 meters). And each tree may be home to thousands of other plants that live on it, each getting its water from the wet air, their roots hanging from the green canopy above. Such plants, known as **epiphytes**, may cover the trees. Other plants, known as **lianas**, form woody vines that climb up the tree trunks like snakes. Animals, too, are abundant and varied in tropical rainforests. These are lands of extreme **biodiversity**. Millions of species are known to exist there; millions more that are as yet to be discovered may exist there as well.

In places where the mountains are too high for the clouds to climb over them, the trees intercept the airborne moisture (Figure 1-1). Condensation forms on the trees and drips to the ground. In such cloud forests, the amount of moisture delivered to the ground by fog and clouds may exceed that delivered by rain.

Where the mountains are really tall, all of the moisture carried by the air may be wrung out of it, and as the air begins descending the west side, it may

Figure 1-1 Mountain enshrouded with clouds.

come down dry. In the lee of such mountains, one may find another kind of tropical forest, the dry tropical forest. In other places, one finds deserts.

Tropical rainforests cover approximately 6% of the Earth's surface. A hundred years ago, they may have once covered as much as 15%, but human activities have reduced them severely, and the destruction continues.

As the name implies, tropical rainforests are found in the tropics, almost in a belt circling the globe around the equator (Figure 1-2). The largest tropical rainforest is found in northern South America, largely in Brazil but also up the Atlantic coast to include French Guiana, Surinam, and Guyana. It also extends west into the foothills of the Andes in Peru, Ecuador, and Colombia. A separate stretch of rainforest occurs in western Colombia and continues north through Panama, Costa Rica, Nicaragua, El Salvador, Honduras, Guatemala, Belize, and into southern Mexico. The second largest rainforest is in Africa beginning in Guinea on the Atlantic coast and continuing south and east along the shore of the Gulf of Guinea as far south as Gabon. It continues east into Congo and Zaire and beyond the Rift Valley. On the Pacific side, the island of Madagascar off the east coast of the continent was once entirely forested. Today, much of that forest is gone, but the fraction that remains continues to hold plants and animals found nowhere else.

Finally, in Asia and Australia, tropical rainforest extends from northeastern India around the Bay of Bengal and down the Malay Peninsula, through Indonesia and the East Indies to include the island of Borneo and along the east coast of Australia from the Cape York Peninsula south toward

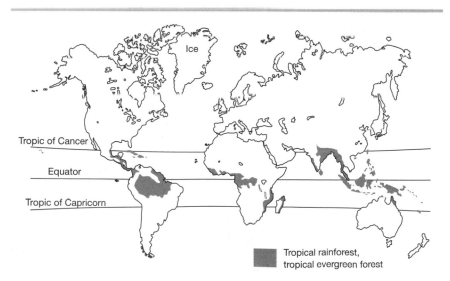

Figure 1-2 World map showing tropical forests.

the Gold Coast. Another region stretches from the island of Borneo, through the Philippines, and north through Viet Nam.

Another type of rainforest exists in the North Temperate Zone, but it is substantially different and more appropriate to another book.

Tropical rainforests have been described as the lungs of the planet. The year-round photosynthesis carried on by the tropical plants takes carbon dioxide out of the air and replaces it with oxygen. In fact, it is thought by some that the destruction of the rainforests is one of the principal causes of the increase of carbon dioxide in the atmosphere and the resulting global warming. More important than the loss of their contribution to the global carbon balance, however, is the loss of their biodiversity. Medically important plants, such as the sources of the drugs quinine and reserpine, are from the rainforest. Some of the as yet undescribed species that may exist there could conceivably hold the cures for cancer or Parkinson's disease, or there may be other kinds of resources as yet unimagined. At the very least, the ancestors of many of our domestic plants and animals have their origins in the tropical rainforests. Relatives of those organisms still live there and serve as genetic reservoirs. If the tropical rainforests disappear, the gene reserves to turn to when they are needed will no longer be there.

An Overview of the Tropical Rainforest

To many, the idea of a tropical rainforest is a dense, perhaps impenetrable tangle of vegetation that is crawling with poisonous snakes, alive with clouds of blood-thirsty mosquitoes, and housing large, man-eating animals as well. Admittedly, all of those do exist, but they're the exception rather than the rule. Most people who visit the rainforest never encounter snakes and man-eaters, and the mosquitoes are usually much less dense than one would expect. Moreover, a mature rainforest is surprisingly open.

The jungle concept originated with European explorers whose first encounter with the tropical rainforest was most likely the dense mangrove swamps that grew along the coasts and tidal rivers. There, light is available to the plants and they respond with dense growth. The Europeans inferred that the entire tropical forest was like that. In reality, however, the rainforest appears to be stratified or arranged in three layers: the **canopy**, the **understory**, and the **forest floor**.

The canopy, as its name implies, is the uppermost layer. It consists of the crowns of the tallest trees and the plants and animals that live in them. Technically, there is a higher layer: the **emergents**. These are the tallest trees that tower above the rest, perhaps to heights of as much as 200 feet, as is the case with the Kapok tree of the American rainforests. Emergent trees, whose tops may more or less resemble broccoli florets, are widely spaced. However, approximately 80 feet above the ground is the canopy where the bulk of rainforest activity occurs. Here one finds the ends of branches of trees overlapping and intertwining, and the branches themselves are covered with small, herbaceous plants known as epiphytes (Figure 1-3). These plants grow in the canopy because that is where the light happens to be. The dense canopy shades everything that grows beneath it. The epiphytes get moisture from the humid air and mineral nutrients from the airborne dust. Their roots hang from the branches toward the ground and sometimes to the ground. In addition to light, there is abundant air movement in the canopy, and moisture evaporates readily out of the trees.

A number of plant families can be found among the epiphytes, including cacti. In the forests of Belize, for example, cactus is a common epiphyte. Normally found in deserts, cacti can live in the humid rainforest up in the tops of trees where they get the small amount of moisture they need from the air.

Perhaps the most interesting of the epiphytes is the family of plants that includes the pineapples: the bromeliads. These are plants whose long, narrow leaves grow in concentric rings. Between each ring, water can accumulate, and within the water one may find mosquito larvae, tiny polliwogs of tree

Figure 1-3 Rainforest canopy. Note epiphytes growing on tree branches.

frogs, and any number of other kinds of small aquatic animals. Canopy-dwelling animals like monkeys and birds may drink out of the bromeliads.

Another plant family that is well represented among the epiphytes is the orchids. (Figure 1-4). One of the largest plant families in the world and common over most of the planet, the orchids are most abundant and diverse in the tropical rainforests. Their blossoms vary from big and gaudy to minute, and they employ all manner of animals to pollinate them. Some even blossom at night when they can attract bats and night-flying moths.

Animals abound in the rainforest canopy too. **Arboreal** monkeys and apes swing from limb to limb, occasionally stopping to eat ripe fruit or leaves. In the rainforests of Sumatra and Borneo, canopy inhabitants include the orangutan, an animal that may weigh over 100 pounds. Other climbing mammals may scamper through the branches in search of food or in flight from a predator, while predators like jungle cats search for prey. And, of course, bats abound at night. Some of them dart among the branches for insects while others sip nectar from night-blossoming flowers. Birds, many of them brightly colored, also dart from tree to tree. Many of the birds that spend summers in the northern hemisphere migrate to the rainforest for the winter, such as Baltimore orioles and a number of warblers of the United States. Bird diversity,

Figure 1-4 Orchid.

however, is extensive in the tropics: Fifty-one hummingbird species can be found in Costa Rica in contrast to only four north of Mexico.

Snakes also slither through the rainforest canopy. Many, like boa constrictors, may be arboreal when they're young, but they largely become forest floor dwellers once they've grown to full size. Others, like the emerald tree boa, spend their entire lives in the trees. Amphibians also live in the canopy (Figure 1-5). Some types of frogs lay their eggs in bromeliads; others bear their young alive as miniature adults, totally skipping the tadpole stage. In contrast, salamanders are scarce.

The most abundant animals in the canopy are the insects. Perhaps the most noticeable are the brightly colored butterflies, but many other kinds are abundant as well, including ants, beetles, flies, bees, and mantises. Ants, in particular, are extremely varied, and some live in trees, which they protect from invaders. There are a variety of spiders in the canopy as well.

Beneath the canopy of the tropical rainforest is the understory. This is a well-shaded area in which there is little air movement but high humidity. There is also very little sunlight. The plants that make up the understory are usually the young of canopy species. They may remain small for years, but as soon as light enters their world, they shoot up. Light may enter the understory when a mature tree falls. In fact, whenever a mature tree falls, many smaller plants take advantage of the temporary window and grow. Sooner or later, however, one or a few of them will dominate and begin shading the others. Eventually, the hole in the canopy closes and the understory returns to relative darkness.

Although there are no animals unique to the understory, it is visited by those from both the canopy and the forest floor. Monkeys from the canopy as well as other arboreal animals may descend into the understory, and ground-dwelling animals, such as chimpanzees in Africa, may climb up from beneath. Some animals may pass through the understory from the forest floor to the canopy. In the rainforests of Central and South America, for example, bullet ants and leaf cutter ants that live in the ground at the bottoms of trees climb the trees to get to the canopy.

Figure 1-5 Tropical tree frog.

The bottom layer of the rainforest is, of course, the ground, or the forest floor as it is known. It is usually covered with a thin layer of leaves and other debris from above, and it has its own collection of unique residents. There are few plants that are limited to the forest floor. Too little light manages to penetrate the canopy to support low plants, but there are bacteria and fungi that live on the forest floor that decompose the litter that accumulates there. In addition, there is an abundance of animals, from small insects to larger animals like deer in the Americas to gorillas in Africa to hunters like tigers in Asia.

The Kinds of Rainforests

People who live in the North Temperate Zone, the bulk of the populations of North America and Europe, generally are unaware of the variety of rainforests that exist. Indeed, many people would not be able to tell the difference between different kinds of forests even if they saw them. After all, many of them do look much alike. However, rainforests occur in a number of different geological and even climatological circumstances. It only makes sense that there would be differences among them.

Equatorial Rainforest

Perhaps the typical rainforest, if such a thing exists, is illustrated by the equatorial rainforest, as is found in the Amazon basin in Brazil (Figure 1-6). There the temperature is characteristically between 25°C and 35°C (75°F to 95°F), and the rainfall varies between 60 to 150 inches (1500 to 4000 mm) per year. There is little seasonal variation in either temperature or rainfall, and the trees keep their leaves all year. Emergent trees in the equatorial may reach heights of 200 feet or over 60 meters.

The Brazilian rainforest along the Amazon basin is truly unique. Although it does not experience seasonal changes in temperature, the source of the Amazon in the Andes does. As the mountain snow melts in the spring, the river rises flooding the forest around it. As a result, the animals of the forest floor are forced to retreat from the rising water, and animals of the canopy may have to swim if they wish to get to a different tree. Fish swim around the trunks, as do the Amazon River dolphins, an unusual species that lives in fresh water. Some of the fish are actually essential to the perpetuation of the forest. Some of the trees produce seeds that cannot germinate unless they first pass through the gut of certain fish. The digestive system of the fish softens or even cracks the tough covering of the seed but does not digest it and passes it out of its body when it defecates. The young plant can then penetrate the seed coat and sprout once the floodwaters recede.

Figure 1-6 Equatorial rainforest.

Subtropical Rainforest

North and south of the equatorial rainforest are the subtropical rainforests. To the untrained eye, these are indistinguishable from equatorial rainforests, but they do contain different species of plants and, to a large extent, animals as well. In addition, they also show some seasonal variation, although, admittedly, not much, especially where they are close to the equatorial forest. However, there is a dry, or perhaps more appropriately a drier season, and one does see some temperature fluctuation, especially as one gets farther from the equator. These forests extend to as much as 10° north and south of the equator.

Cloud Forest

Both the equatorial and subtropical rainforests might be considered lowland forests. They generally grow close to sea level. Along the coasts, they often have coconut palms that are absent inland, but conditions change as one gets into higher elevations. Generally above 1000 meters the air becomes cooler than at sea level, especially at night, and much of the moisture in the air begins to condense. At higher elevations there may be condensation throughout the day. This is the cloud forest. Much of the moisture in the mist that forms accumulates on the trees and leaves, and a great deal of it drips down to the ground. Palms in general are distinctly absent, but other plants that are often not encountered in lowland forests are present. In the cloud forest, ferns grow to the size of trees. Such tree ferns can be described as living fossils; they date back before the dinosaurs. In addition, the trunks of the trees are often covered with mosses and liverworts, primitive plants that thrive in darker, wetter conditions. In contrast, in lowland forests, one sometimes finds the trunks of trees bare.

Much of the water in the cloud forest is essentially recycled from the condensate that drips to the ground. Plants soak up more water than they can use, and the excess evaporates out of the leaves, a process known as transpiration. From a distance rising columns of vapor almost look like smoke, as if the mountains are on fire. The transpirational water combines with the cloud water to feed the forest. This process is not unique to the cloud forest; it occurs in other rainforests as well, although possibly to a lesser extent. Consequently, when rainforests are cut down, the recycling of rain water stops, and everything that falls as rain ends up running off the land and into the rivers. If too much of the rainforest is destroyed, it may not recover.

There are some animals that are unique to the cloud forest as well, and some are rare, even endangered. In the cloud forests of Costa Rica in Central

America, for example, the resplendent quetzal is seen nowhere else in the country. Coincidentally, the quetzal is the national bird of Guatemala where the cloud forest is abundant. The mountain gorilla is another unique cloud forest animal. It lives in equatorial Africa in the mountains of eastern Zaire and western Uganda at elevations up to 3900 meters.

Monsoon Forest

The last of the true tropical rainforests is the monsoon forest, which is found for the most part in Asia in the northeast region of India and around the northern shore of the Bay of Bengal. It extends north of the subtropical forests, and it is characterized by a distinct dry season. During the dry season, trees lose their leaves just as they do during the winter in the Northern Hemisphere, and fires become a distinct possibility. In contrast, during the rainy season, winds blow in from the oceans laden with moisture, and the rains may be torrential. In Bangladesh, for example, runoff from monsoon rains can swell rivers and cause extensive flooding, particularly close to the coast.

Mangrove Swamp

The final tropical ecosystem that is dominated by trees is technically not a forest; it is the mangrove swamps that are found along quiet shorelines in the **tropics**, the region of the planet between the Tropics of Capricorn and Cancer (Figure 1-7). Mangroves are halophytes: plants that grow in salty environments. This includes not only the shorelines but also along the edges of river estuaries perhaps as far upstream as the rivers are influenced by tides. Mangrove trees produce dozens of adventitious roots, which originate high on the trunks or on the branches. These form tangles of complex stilt-like supports, much of which are underwater during high tide and some of which are constantly underwater. The roots slow down the movement of the water, both tidal movement and the current of rivers, causing sediment that the moving water may be carrying to settle out. In addition, the roots often catch floating material such as logs and other kind of organic debris. The accumulation of these materials builds land, which explains the extensive development of deltas at the mouths of rivers such as the Amazon and the Mekong. More recently, the accumulating debris has included plastic bags, foam cups, plastic bottles, aluminum cans, and other kinds of litter that is either carried downstream by rivers and/or has been tossed overboard from ships.

Mangrove swamps play other important roles in nature. The tangle of roots forms underwater labyrinths that offer abundant hiding places for small fish and arthropods. The arthropods may include shrimp, and the small fish

Figure 1-7 Mangrove swamp.

may be the young of commercially important species. Moreover, the underwater parts of the roots are covered with colonies of sessile animals, many of which are stinging hydroids, like tiny sea anemones. Thus the small fish and arthropods are not only able to hide among them, they're also protected from larger predators by what amounts to living barbed wire.

Perhaps the most important role of mangrove swamps from a human point of view is that they absorb storm surges. When a tropical storm or hurricane develops, the winds spin in a counter-clockwise direction around the center of the storm or eye. This generally causes the sea water to pile up on one side of the eye, usually toward the north or east, as the storm approaches land. The stronger the wind, the more the water will pile up. This is the storm surge, and it hits land in a series of giant waves. Where the shore is protected by mangrove swamps, the storm surge generally does little damage. However, where the swamps have been removed for development, damage can be extensive. Once the Gulf Coast of the United States was extensively lined by coastal swamps, as was much of the Yucatan peninsula of Mexico. However, as these coasts were developed, swamps were cut down and homes, golf courses, and resorts were built, as were oil refineries and shipping ports. Consequently, periodic hurricanes and tropical storms have caused havoc. One example, though oddly not in a tropical forest, is the destruction of the city of New Orleans by Hurricane Katrina in the autumn of 2005. It illustrates the kind of damage that such storms can do. Extensive damage also occurred in Biloxi, and Gulfport, Mississippi. In 2004, three hurricanes did major damage along the gulf coast of Florida, and Hurricane Wilma caused extensive damage in the Yucatan in 2005. This is not to say that mangrove swamps would have prevented all of that damage, but it is highly likely that they would have substantially reduced it.

Like the mangrove swamps, the tropical rainforests have experienced extensive destruction, and that destruction affects the human race, directly and indirectly. That destruction will be discussed later in this book. However, the rainforest represents an enormous resource and a crucial ecosystem in the overall conservation of the planet. It is important to understand that resource because understanding leads to preservation, and with all the potential the rainforest holds for making human life better, its preservation is important.

The Tropical Water Cycle

The Origin of the Rain

AMONG THE THINGS THAT CAN strike someone when he or she first enters a tropical rainforest are the lush vegetation, the humidity, and the heat. The first two are factors of the rain the forest receives, which may also be striking if it happens to be falling during one's entrance. Although it rains a lot there, it doesn't rain constantly. Rainfall varies over the day and over the year. Rainfall in the tropics is largely a factor of moisture being carried from the ocean by the wind, and the wind there comes principally from the east.

Wind direction results from the interaction between air movement and the rotation of the Earth. The intense sunlight that strikes the Earth in the tropics, especially around the equator, warms the air, and the warm air begins to rise. This is enhanced by the northeast and southeast trade winds converging upon each other near the equator, essentially forcing the equatorial air to rise. The area where this happens is known as the **Intertropical Convergence Zone** (ITCZ).[1] As it rises, it encounters less and less atmospheric pressure and expands, and as it expands, it cools. The cooler air can hold less moisture than warm air; consequently, airborne moisture condenses as the air rises and cools, and clouds form. As the air continues to rise, it loses its ability to hold moisture entirely; the moisture condenses still more until it condenses to water, which falls as rain. The higher the air rises, the cooler it gets and the more moisture it loses. Eventually, perhaps as it reaches the upper limits of the atmosphere, it dries and begins spreading out toward the north and south.

Once the now dry air reaches approximately 30° north and south of the equator, a few degrees beyond the latitudes of the Tropics of Cancer and Capricorn respectively, it mixes with air moving in from the arctic and antarctic circles, and it sinks. As it descends deeper and deeper into the atmosphere, it encounters greater air pressure and contracts. As it contracts, it warms and dries even more. Eventually, it hits the Earth's surface, hot and dry. Where this occurs over land, we find some of the world's greatest desserts.

As the sinking air hits the rotating Earth, it receives an acceleration toward the east, the direction of the Earth's rotation, proportional to the planet's speed at its point of contact, a bit less than 800 mph (1280 kph). Because the Earth becomes progressively more narrow as one travels toward its poles, the speed at which a point on its surface is moving would be slower at the arctic and antarctic circles than it would be at the Tropics of Cancer and Capricorn. In contrast, the Earth is at its widest at the equator. A point on the equator would move at a greater speed than one anywhere else on the planet's surface, greater than 1000 mph (1600 kph). Thus, air striking the Earth at the Tropic of Cancer would receive an acceleration of approximately 800 mph (1300 kph), as mentioned above. If it moves toward the north, it would travel over a progressively slowing surface and drift toward the east. Air moving south from the Tropic of Cancer would, conversely, drift toward the west because it would encounter a progressively faster surface. This, then, results in the northeast trade winds that occur in the tropics north of the equator. In the Southern Hemisphere, what happens mirrors the events in the north. As a result of this, the prevailing direction of the winds in the North and South Temperate Zones is towards the east, while in the tropics it is *from* the east, and most rainforests are situated on the eastern sides of continents and tropical islands. This, admittedly dry, description of air circulation on the Earth's surface and the resulting winds is a result of what's known as the **Coriolis effect**, illustrated in Figure 2-1. Although surface features such as mountains can exert a regional influence on wind direction, the Coriolis effect accounts for the prevailing direction of the winds over most of the planet.

Warm air that travels over open ocean picks up moisture in the form of water vapor. The longer it is over the ocean, the more vapor it picks up; in time it can become saturated. If it encounters land before it has lost its moisture, things begin to get complicated. Land surfaces in the tropics are often warmer than the water. Consequently, saturated air will rise over land, airborne water vapor condenses to rain, and the lush plant growth typical of the tropics is one of the results.

Within the Forest

When rain hits the ground, it can either run off the surface directly back to the ocean or into freshwater bodies, or it can seep into the ground. Three things can happen if it seeps into the ground: it can run underground back to the ocean or into freshwater bodies, it can become stored underground in **aquifers** as **groundwater**, or it can be absorbed by plants and used in the biochemical process of **photosynthesis**, the way plants use water and atmospheric carbon dioxide to make glucose. Glucose is a simple sugar that can be used directly, usually for energy, or converted to other compounds. Virtually all compounds used by plants and animals are ultimately derived from glucose. Photosynthesis is usually represented by the equation

$$6CO_2 + 6H_2O \Rightarrow C_6H_{12}O_6 + 6O_2$$

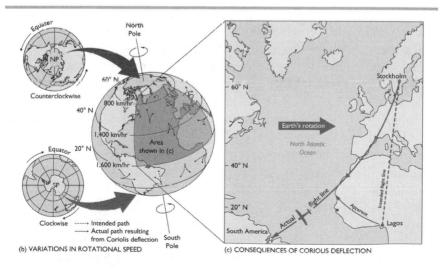

Figure 2-1 Coriolis effect.

where CO_2 represents carbon dioxide, H_2O represents water, $C_6H_{12}O_6$ represents glucose, and O_2 represents oxygen. The actual process of photosynthesis occurs usually in the leaves of green plants, and it is far more complex than the simple equation above implies. The plant splits the water molecule; it combines the hydrogen with the carbon dioxide to make the glucose, and it releases the oxygen into the air. Indeed, the oxygen that many forms of life, ourselves included, need in order to survive comes from the photosynthesis of green plants. However, not all of the water the plants absorb, particularly in the tropics, is used in photosynthesis. Some may be simply pushed out of the leaves, especially when there's too much water present to be used, a process known as **guttation**, and some water will evaporate from the leaves, a process called **transpiration**. Guttational water will often be recycled through the soil and back to the plant, but water lost by transpiration combines with moisture already in the air. Sometimes it condenses almost as quickly as it leaves the plant. One can sometimes see dense clouds of vapor rising from the rainforest like smoke rising from a fire. Much of this moisture is returned to the forest. In fact, when rainforests are **clear-cut**—that is, all trees are removed from a substantial region—rainfall diminishes. Moreover, more water from what rain there is runs off the ground surface, often carrying soil particles with it, and the ground, which is usually pretty boggy, dries as water evaporates from it. Because clear-cutting removes the trees that slowed runoff, held the soil in place, and pumped enormous quantities of moisture back into the air, what was once lush rainforest becomes arid grassland. Some scientists believe that the Mayan civilization (in what is now Guatemala, Belize, and tropical Mexico) may have collapsed as a result of deforestation. The Mayans cleared much of the rainforest in which they lived, and the ensuing lack of rainfall brought on drought and crop failure. In summary, then, within the greater scheme of circulation of water on the planet, or the **hydrologic cycle** (Figure 2-2) as it is called, there are at least two subcycles in the tropical rainforest that are necessary to keep the greater cycle going, and they both absolutely require the trees.

Other factors are involved in the delivery of rainwater to the tropical forests. Sometimes, usually at night, air temperature can drop sufficiently for moisture in the air to condense. The resulting fog droplets cling to whatever vegetation they encounter, run down tree trunks, and drip from leaves. A considerable amount of water is delivered to the ground and, subsequently, to the plants this way. Additionally, saturated air that is forced to climb mountains will undergo the same events. This, as described in Chapter 1, can result in a unique form of tropical rainforest known as cloud forest.

My own experience with the cloud forest was in the Monte Verdi region of Costa Rica in Central America, which I visited twice. I did not see the sun

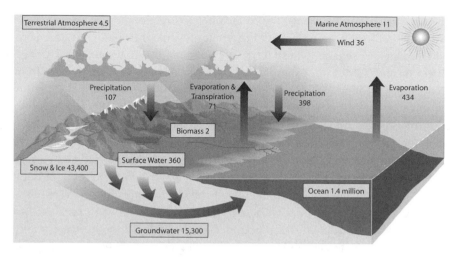

Figure 2-2 The hydrologic cycle.

once in either visit, and I managed to saturate every item of clothing I wore that wasn't covered by raingear simply by hiking through the forest. Moreover, the clothing did not dry until I left the region. I was frequently awakened at night by the sound of rain pelting the roof of my cabin, and a pair of boots that belonged to a colleague grew an intriguing green mold. In spite of the discomforts, however, I would return at my first chance, simply to see the variety of life that exists in few other places, although I'd have to travel through much depressing pastureland cleared from the forest. More than 50 species of hummingbirds live in the cloud forest, along with resplendent quetzals and hundreds of other beautiful birds. Every tree seems to be completely encrusted with epiphytes, plants that grow on other plants, described more fully in Chapter 5, their roots dangling in the air soaking up the moisture. Unfortunately, perhaps, many people must feel as I do; Monte Verdi has become a haven for tourists. In fact, one Costa Rican guide I know now refuses to take his clients to Monte Verdi because of its popularity. He feels that its original nature has been compromised, and he refers to it as the "crowd forest."

Seasonal Changes

The general perception of a rainforest is that rain falls more or less evenly throughout the year. In much of the equatorial rainforest, this may be the case. However, in the northern and southern extremes of the biome, there may be seasonal variation. In Central America, for example, there appears to be a distinct rainy season from May to October, corresponding more or less to summer in the Northern Hemisphere, while November through April are considered the dry season. These terms are not absolute, however. In the Congo River basin of Africa, for example, the so-called dry season usually lasts from November through March, comparable to that in Central America. The difference between it and the rainy season is that it simply rains less. Even in Central America, the seasons are not carved in stone. I was in Belize in northern Central America once during August, the height of the rainy season, supposedly, and a drought was in progress, to the point that we had to conserve water. Showers were forbidden, and drinking water had to be brought in by truck. The good news was that the usually ubiquitous mosquitoes were missing. In contrast, I was once at a remote camp in Caribbean Costa Rica in January where a tropical storm pounded us for days. When we were finally able to hike back to civilization, we had to slog through knee-deep mud and wade up to our waists or higher through swollen streams. When we arrived, we learned that roads and bridges had been washed out, the boat that was carrying our gear capsized in the surf, and some of the crewmen drowned. This during the dry season!

The meteorological forces driving the rainy and dry seasons may be identical, at least in part, to those driving the cold and warm seasons in the Temperate Zones. During the Northern Hemisphere summer, the North Pole is tilted toward the sun, and the northern half of the Tropical Zone receives intense sunlight. Ocean water warms, evaporates more readily, and the conditions described at the beginning of this chapter occur, with rain getting dumped abundantly on tropical lands. In contrast, during the northern winter, the north of the Earth is tilted away from the sun, the northern tropics receive less intense sunlight, the oceans cool somewhat, and rainfall is reduced (Figure 2-3). These conditions are reversed in the Southern Hemisphere, of course. In northeastern Australia, for example, around Cairns, the dry season lasts from May through October corresponding to the Southern Hemisphere winter.

An extreme example of alternating wet and rainy seasons occurs in what are known as **monsoon climates**, where there is abundant rainfall, but it's concentrated in a few months. The remainder of the year is dry. Technically, the monsoon is a change in wind direction, and it is perhaps best illustrated in Asia.

According to Michael Ritter,[2] during Asia's northern summer, increasing ground temperatures generate air uplift, generally somewhere over Mongolia. The rising air causes low pressure that draws air in from the surrounding areas as far away as the Pacific and Indian oceans. As the oceanic air crosses onto the continent, it loses moisture, often dramatically. The rain may fall in torrents, and local flooding can occur. During the northern winter, cold ground temperatures cause a reverse in airflow, and the air moves from the continent out over the ocean, generally causing dry conditions. Monsoon climates also occur in Southwestern Africa and in South America around northeastern and southeastern Brazil.

Among the most dramatic examples of monsoon climates is that of Bangladesh, a small country to the north of the Bay of Bengal between India and Myanmar. Three rivers, the Brahmaputra, Ganges, and Meghna, converge there. The monsoon winds that come off the Indian Ocean during the rainy season cross the country and encounter the Himalayan Mountains, which they must climb, losing most of their moisture in the process. As a result, most of Bangladesh receives at least 90 inches (230 cm) of rain per year; some parts receive more. Roughly 80% of this falls during the monsoon season, causing the rivers to rise. In combination with heavy storms and flood tides, the monsoon rains cause disastrous flooding in the river delta region. Property damage and loss of life are regular events.[3]

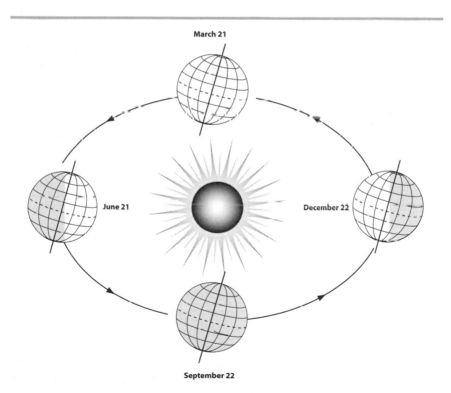

Figure 2-3 Seasons in the Northern Hemisphere. Data courtesy of NASA.

Flooding

Flooding is always an issue when a large amount of rain falls in a short period of time. It's also a concern when a large volume of snow melts. At first, snow might seem like an odd thing to mention in a book about tropical forests. However, the largest tropical rainforest in the world, that of equatorial Brazil, is dominated by the Amazon River, the world's largest river, at least in terms of water flow. The origin of the Amazon is in the Andes, and much of the water that flows down the Amazon begins as Andean snow. The melting of that snow contributes to some of the most impressive flooding on the planet and has given rise to a unique ecosystem: the floodplain forest.

Flooding in the Amazon basin is a more complicated series of events than it is in the rivers of North America. The Amazon watershed covers almost 2.37 million square miles (more than six million square kilometers). It occupies most of Brazil and reaches into Venezuela, Columbia, Ecuador, Peru, and Bolivia. It overlaps the equator, which means that the rainy and dry seasons differ in the northern and southern extremes of the watershed. Moreover, in addition to rainforest, the climates of the watershed include monsoon and, in the case of the high Andes, snowy winter with spring melt. Consequently, there is no one time of the year that can be described as the flood season throughout the entire watershed.[4] However, the true flooded forest occupies only a fraction of the region, about 335,000 square miles, in northern Central Brazil, and the flooding occurs from June to October,[5] although other parts of the basin will flood at other times for different durations. During the flooding, water level may rise as much as 30 feet (9 m), inundating the surrounding forest and covering the soil and lower trunks of trees. Essentially, what had been forest becomes swamp.

As would seem obvious, flooding has an enormous impact on the surrounding area. For one thing, the annual cycle of flooding deposits silt and plant nutrients in the soil, enriching it. In addition, the flooding process will dig up areas where water remains when the flood recedes. These lakes provide habitat for animals, and probably plants as well, which do not prosper as well in the open river. In addition, the resulting lakes may also provide nursery areas for some species of fish and perhaps for animals like caimans. When these lakes are inundated during the following flood season, the animals would be able to migrate into the main river.

The plants and animals of the floodplain forest are well adapted to this environment. Fishes readily forage among the flooded tree trunks, as do pink river dolphins, one of two species of freshwater dolphins found in the Amazon basin. Of particular interest is the relationship some species of fish have developed with the forest plants. While we tend to imagine things such as piranhas and electric eels when someone mentions the Amazon, there are hundreds of other species as well, and some are strict vegetarians. Some fish even eat fruit. When the forest floods, certain plants drop their fruit to be eaten by fish. The fish feast on the fruit and either release seeds to be carried away and deposited elsewhere by the water, or they swallow the seeds, which pass unharmed through the fish's digestive system. When the fish defecates, the seeds are released. If the seeds land somewhere that will be dry once the water recedes, it can become a new plant. Other seeds actually require that a fish eat them. Such seeds must pass through the fish's gut in order to have a hard covering either softened or removed, or they wouldn't otherwise germinate.[6]

Despite its location in a region of continuous warm weather, the tropical rainforest still can show some seasonality, whether it's a factor of locally fluctuating rainfall or true seasons somewhere else, as is the case of the Amazon and the Andes. Whatever the cause, the indigenous plants and animals have adapted to it, as have the indigenous people. It's a lesson we'd do well to learn in North America.

References

1. International Research Institute for Climate and Society. 2007. *Understanding the ITCZ*. http://iri.columbia.edu/ ~ bgordon/ITCZ.html.
2. Ritter, M. E. 2006. *The physical environment: An introduction to physical geography*. http://www.uwsp.edu/geo/faculty/ritter/geog101/textbook/title_page.html.4.
3. Library of Congress. 2005. *A country study: Bangladesh*. http://lcweb2.loc.gov/frd/cs/bdtoc.html#bd0035.
4. World Resources Institute. *EarthTrends: The environmental information portal*. http://earthtrends.wri.org/maps_spatial/maps_detail_static.php?map_select = 410&theme = 2.
5. WideWorld Ecoregion Profile. *Amazon river and flooded forests*. http://www.nationalgeographic.com/wildworld/profiles/g200/g147.html.
6. Kricher, J. 1997. *A neotropical companion*. Princeton, NJ: Princeton University Press.

Forest Soils and Nutrient Cycles

3

Soils

WHEN I WAS A STUDENT taking a course in evolution, the professor told of a time he received a package containing only a ball of dried clay and instructions to place the clay in an aquarium overnight. He did so, and upon his return to the lab the following morning, he found the clay scattered over the bottom of the aquarium and a fish resting on top of it.

Instant fish: simply add water.

The fish in question was a lungfish, a member of a group of primitive freshwater fishes that are capable of breathing air and today are found only in Australia, South Africa, and South America. The likelihood that such similar fish would originate and evolve independently in three such disparate locations is remote. A more scientifically plausible explanation is that the three current populations are remnants of what was once a more widely distributed one, and that the three now separate areas were once joined together.

The locations of the world's landmasses have not always been as they now are. Indeed, since the formation of the planet, the continents have drifted around, sometimes connected to one another and sometimes not. In fact, around 200 million years ago, all of the land masses were combined into a single super continent called Pangea (Figure 3-1). The southern portion of this land mass, which contained the southern-most continents and India, was called **Gondwanaland.** If you look at the east coast of South America and the west coast of Africa, you can almost see how the two continents were united. Around 170 million years ago, all of the continents in the Southern Hemisphere as well as India were joined in a super-continent known as Gondwanaland (Figure 3-1). Today, most of the world's great tropical rainforests occur on the Gondwanaland continents. That these land areas were once grouped together can explain how, though now distant from one another, they can contain some similar animals, such as lungfish, and it can partly explain why the major rainforests have similar soils.

The phenomenon behind the movement of the continents is known as **plate tectonics.** To understand this, picture the surface of the Earth divided into panels, more or less like the surface of a soccer ball, although unequal in size. These panels, or plates, float on the molten **mantel** of the Earth. Most of the plates are covered by the oceans; landmasses represent areas of plates that have been forced upward. Where two plates abut against each, for example the San Andreas Fault of California, violent geologic forces such as earthquakes or volcanoes occur.

In general, soils are made of three basic particles: sand, silt, and clay. Clay particles are the smallest of the three; in fact, they're small enough to allow the electrons of their atoms to cause a slight negative charge on the particle's surface. Cations, such as those of potassium or calcium, are attracted to the particles and cling loosely.

Rainforest soils are dominated by clay particles. Although one finds local differences among the soils, that they are overwhelmingly similar supports the idea of their common origin.

Soil is a very dynamic medium. It is the site of many physical and chemical events, and it supports a complex biota as well. Soil seems to naturally stratify into four distinct layers or **horizons** (Figure 3-2): topsoil (A horizon), subsoil (B horizon), parent material (C horizon), and bedrock (D horizon). The bedrock, as its name implies, is an underlying layer of rock, pieces of which periodically break off to form the parent material. Physical and chemical events, collectively known as **weathering**, break down the parent material to form the subsoil. Tropical rainforest soils are highly weathered. Soils of wet, low-lying areas

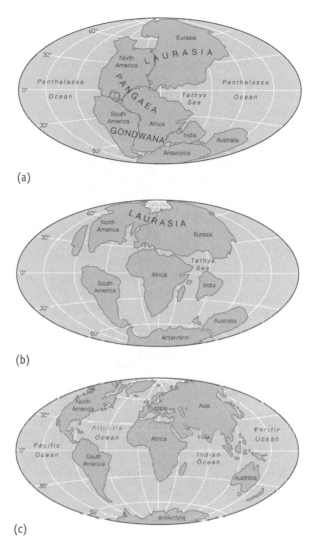

(a)

(b)

(c)

Figure 3-1 Gondwanaland showing modern continents conjoined.

generally tend to be yellow-orange in color, and those in drier, upland regions are often reddish. These colors are due to iron compounds.[1]

The topsoil originates from the mixing of mineral material from the subsoil and the decomposition products of dead plants and animals. Thus, the topsoil is where one finds organic material plus many inorganic nutrients. In some ecosystems, dead biological material such as leaves and animal droppings accumulate in a distinct layer above the topsoil. This is known as **litter**.

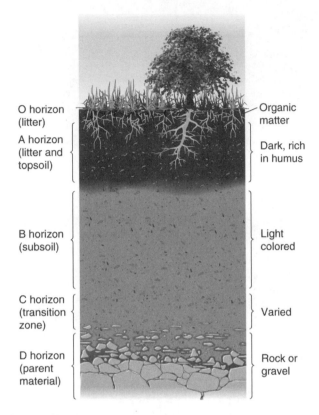

O horizon (litter) — Organic matter

A horizon (litter and topsoil) — Dark, rich in humus

B horizon (subsoil) — Light colored

C horizon (transition zone) — Varied

D horizon (parent material) — Rock or gravel

Figure 3-2 Soil profile diagram.

In some ecosystems, decomposed litter may form a horizon of its own between the nondecayed litter and topsoil.

Tropical rainforest soils are characteristically composed of clay, and they have little or no topsoil. They also have a very thin litter. Because there is neither a dormant nor totally dry season in a tropical rainforest, monsoon forests being the exception, any plant or animal material that falls to the forest floor undergoes decomposition quite rapidly. Organic material and mineral nutrients that find their way into the soil are quickly recycled into the trees, where much of the rainforest's nutrients actually accumulate. Moreover, the abundant rainfall carries away most of the mineral cations that may have attached to the clay particles, a process known as **leaching**. Furthermore, when rainwater reacts with carbon dioxide in the soil, some of it undergoes the reaction

$$H_2O + CO_2 \Rightarrow H_2CO_3$$

to form carbonic acid. The carbonic acid then dissociates to a hydrogen ion and bicarbonate

$$H_2CO_3 \Rightarrow H^+ + HCO_3^-$$

The bicarbonate routinely is leached from the soil, but the hydrogen ion often attaches to the negatively charged surface of clay particles, sometimes dislodging other cations in the process. As a result, tropical rainforest soils are often acidic.

Tropical rainforest soils are, in general, infertile. Exceptions exist, of course, and some will be discussed later in the chapter. In general, however, rainforest soils are highly leached and weathered clays that are poor in phosphorus, calcium, and potassium,[2] three important mineral nutrients.

It would at first seem paradoxical that the massive trees and extreme biodiversity found in a tropical rainforest would grow and develop on some of the least fertile soils on the planet, but it's easily enough explained. First, the year-round warm temperatures mean that decomposition of anything organic that falls on the forest floor will be rapid. Moreover, plant debris falls at a more or less uniform rate. There is no great leaf fall nor cold winter as there is in a northern forest; therefore, there is no build-up of soil litter. Second, the vast network of soil **mycorrhizas**, soil fungi that form a **mutualistic** relationship with trees, quickly absorb nutrient materials liberated by decomposing vegetation and pass them to the trees. The rainforest is supported by the rapid cycling and recycling of nutrients from and back to the vegetation.

Nutrient Cycles

The tropical rainforest **biome**, like all others on Earth, is described as an open system. This means that it is capable of exchanging both material and energy with surrounding systems. The Earth itself is a closed system: it is capable of exchanging only energy with its surroundings. The minute amount of matter that we import from and export to space is immaterial in terms of the total amount on the planet. However, that the systems on the planet are open is critical for the support and preservation of life. If, for example, all of the chemicals that are tied up in a plant were not returned to the soil following the plant's death, all nutrients eventually would be tied up in a thick layer of litter, and all life would cease. What happens is that chemical nutrients cycle between the living and nonliving worlds through a complex series of events much like the hydrologic or water cycle described in the previous chapter. In fact, the water cycle represents a composite of parts of two separate element cycles: hydrogen and oxygen.

The Nitrogen Cycle

One of the most important plant growth nutrients is the element nitrogen. In fact, nitrogen is the fourth most abundant element in all living things. It is critical for the synthesis of proteins, nucleic acids, and the chemicals involved in respiration and energy processing. It makes up almost 80% of the air in our atmosphere, but despite its abundance, it cannot be obtained directly from the atmosphere by any plant or animal. Instead, it has to be processed or fixed by soil microorganisms before it can be used. The entire flow of nitrogen from the air, through the soil, plants, animals, and eventually back into the air is referred to as the nitrogen cycle (Figure 3-3).

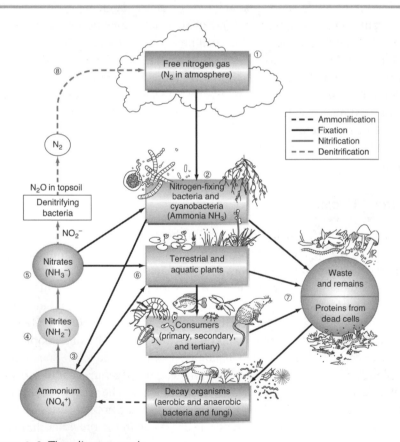

Figure 3-3 The nitrogen cycle.

In the air, nitrogen occurs in molecules containing two atoms each. Some soil microorganisms, including bacteria that live freely in the soil, are capable of taking atmospheric nitrogen and reducing it to ammonia (NH_3), which quickly combines with hydrogen ions and becomes ammonium (NH_4^+). The ammonium may be absorbed as is by plant roots, or it may be oxidized to nitrites (NO_2) or nitrites (NO_3) by soil-nitrifying bacteria. These oxides of nitrogen can then be absorbed by plant roots.

Free-living soil bacteria are by no means the only, or even necessarily the best, nitrogen-fixing organisms. That title most likely belongs to bacteria of the genus *Rhizobium*, which live in a **symbiosis** with plants of the family Fabaceae, the legumes. Legumes are pod-forming plants that include a number of familiar vegetables such as beans and peas, field plants such as clover and alfalfa, and even locust trees and *Saguaro* cactus.

A symbiosis is a relationship where two organisms live together. In the case of the association between legumes and *Rhizobium*, it is a **mutualistic symbiosis** because both members benefit. The *Rhizobium* bacteria benefit by being provided with energy nutrients and other raw materials, while the legume receives fixed nitrogen, a commodity it might otherwise find difficult to obtain. In African and American tropical rainforests, legumes are the dominant plants.[3]

In the rainforests of southeast Asia, legumes are far less abundant than they are in the Americas, and trees of family Dipterocarpaceae are dominant. In these, nitrogen fixation may occur because of **mycorrhizas**. Studies by Kikuchi and Ogawa determined that the ratio of nitrogen-fixing bacteria to non-nitrogen-fixing bacteria were higher in soils where diptocarp mycorrhizas were present than where they were not.[4] In general, trees are far too large to efficiently absorb all of the soil nutrients they require. Symbiotic mycorrhizas penetrate their roots and send filaments out into the soil to absorb water and nutrients, including nitrogen compounds. The volume of mycorrhizal filaments in soils may exceed that of the roots with which they live. In the nitrogen-poor soils of tropical rainforests, the ability of these fungi to support nitrogen-fixing bacteria provides the trees with a badly needed nutrient. The trees, in turn, are more capable of providing the mycorrhizas with the sugars they need.

Another symbiotic nitrogen fixer is the cyanobacterium (blue-green alga) *Anabaena*, some species of which live freely and some of which form a symbiotic relationship with the water fern *Azolla*. Other species of cyanobacteria fix nitrogen too, but symbiotic fixation by *Anabaena* in *Azolla* appears to be particularly productive. This relationship has been used to help fertilize rice paddies in Asia.

The Phosphorus Cycle

A second important soil nutrient is phosphorus, which, like nitrogen, is often the **limiting factor** in plant growth, particularly in the tropics. In the Amazon basin, for example, the ancient soils of the main forest are highly leached and poor in phosphorus. In contrast, soils of floodplains and of the Andean foothills are more fertile in general and richer in phosphorus in particular.[5]

Unlike nitrogen, phosphorus is cycled very slowly, but in many ecosystems, plant roots readily pick up phosphorus that gets into the soil from the decomposition of dead organic material. In tropical rainforests, however, plants will withdraw mineral nutrients, including phosphorus, from leaves before dropping them. Consequently, little phosphorus gets to the soil, and that which does is recycled quickly.[2]

Other sources of phosphorus include the weathering of rock and the deposition of mineral sediments by floodwater. Thus, the greater fertility of the Andean foothills may be a result of the erosion or weathering of the mountains and the deposition of phosphorus-containing material. Similarly, phosphorus dissolved from rock by rain or melting snow may end up being carried into rivers and then deposited on soil when the rivers flood. This results in the constant enrichment of foothill and floodplain soil. Inland forest soil, in contrast, has to make do with whatever phosphorus it can get from the decay of organic material.

Other Nutrients

Other minerals, such as potassium, iron, and sulfur, undergo cycling as well. Some of these can be provided to soil by the decay of organic matter, but some are imported from outside of the ecosystem. For example, the rainforests of Central America are in one of the most volcanically active regions of the Earth. The complete Central American isthmus formed only about three million years ago, practically yesterday in geologic time. Its formation resulted, in part, from the same forces that push the continents around the Earth's surface: plate tectonics. In the case of Central America, two plates, the Cocos to the west and the Caribbean to the east, are pushing against one another forcing the Cocos plate under the Caribbean. The result of this has generated a chain of mountains that contain a string of active volcanoes which, when they erupt, spew out ash containing mineral nutrients that enrich the soils of Central America, making them fertile and productive, despite their being under tropical rainforest. This is why countries such as Costa Rica and Columbia are able to produce agricultural products like coffee and bananas for export.

At one time, people mistakenly believed that the luxuriant growth of the rainforest actually indicated great soil fertility. During the 1960s, for example, it was not uncommon to hear so-called agricultural experts claiming that once we learned how to farm the Amazon, hunger would disappear (the same claim was made for harvesting the oceans).While we now know that neither of these happens to be the case, population pressures and in some cases outright greed are leading to more and more effort to farm the rainforests.

Small-scale attempts at farming the rainforests are often of the slash-and-burn variety. In this practice, a small section of standing forest is cut down, allowed to dry, and burned, often by a small family group. The burning releases nutrients in the plants to the soil, and the plot is then farmed until the soil is depleted. The plot is then abandoned and allowed to recover, while the group moves on to another place. If a small number of people use a large enough area in this manner, no long-term harm is done. However, slash and burn can become problematic where local crowding may force poor people to cut and burn rainforest faster than it can recover. In the country of Uganda, in Africa, for example, slash and burn is not only destroying the forest, it is also threatening rare and endangered animals, such as the mountain gorilla, and it is leading to even greater impoverishment of the people who depend on the forest. By cutting and burning the rainforest, they are depriving themselves of firewood, ironically, and forest resources they use. Once the soil has become too impoverished to grow anything substantial and lays bare, erosion is more likely to occur, as is soil dehydration. Moreover, the carbon dioxide released by the burning and the destruction of the plants that used carbon dioxide in photosynthesis is also contributing to global warming, which threatens all of us.[6]

Despite the fertility of the rainforest soil having been found to be a myth, large-scale agriculture is occurring on rainforest soils. In parts of Central America, vast banana and sugarcane plantations belie that rainforest ever existed there. In the Amazon basin in Brazil one can find even more vast soybean plantations. The Brazilian government had encouraged clearing of the rainforest for farming, and Brazil has become a superpower in soybean exports, second only to the United States. Soybeans are legumes, which means, at least in theory, that they should be providing their own nitrogen to enrich the soil. However, the acid soils of the Amazon basin need to be neutralized with lime, and other fertilizers, plus pesticides and herbicides, must be added as well.[7] Many of these chemicals are lost from where they are applied, carried by wind and rain into rivers and streams, and ultimately to the ocean where they cause ecological damage.

Clearing rainforest for agriculture sets off a series of cascading events that can cause far-reaching change. As mentioned earlier, removing extensive tracts of forest reduces the moisture pumped back into the air by transpiration, which, in turn reduces rainfall causing the ground to dry. Once the ground has dried, typical rainforest plants are no longer able to grow, and grasses move in. Before the grasses can become established, however, wind and whatever rain does fall erode loose soil removing even more fertility. Finally, when the crops are harvested, any nutrients they may have absorbed from the soil are removed with them.

Clearing rainforests, whether for agriculture or any other reason, can be seriously, even permanently, disruptive and will be discussed in a later chapter. The critical point, however, is that the rainforests and the soils upon which they are located are intimately tied to one another. The ultimate consequence of disturbing one is disturbance of the other.

References

1. Schaetzl, R. J., and S. Anderson. 2005. *Soils: Genesis and geomorphology.* New York: Cambridge University Press.
2. Laurance, W. F., S. Bergen, M. A. Cochrane, P. M. Fearnside, P. Delamônica, S. d'Angelo, C. Barber, and T. Fernandes. 2005. The future of the Amazon. In E. Bermingham, C. W. Dick, and C. Moritz, eds., *Tropical rainforests: Past, present, and future.* Chicago: University of Chicago Press.
3. Primack, R., and R. Corlett. 2005. *Tropical rain forests: An ecological and biogeographical comparison.* Malden, MA: Blackwell Science Ltd.
4. Kikuchi, J., and M. Ogawa. 1995. *Nitrogen-fixing (acetylene reducing) activity in the mycorrhizas of dipterocarp seedlings.* http://www.metla.fi/iufro/iufro95abs/d7pap6.htm.
5. Laurance, W. F., S. Bergen, M.A. Cochrane, P.M. Fearnside, P. Delamônica, S. Agra D'Angelo, C. Barber, and T. Fernandes. 2005. The future of the Amazon. In *Tropical rain forests: Past, present, and future,* eds. E. Bermingham, C.W. Dick, and C. Moritz, Chicago: University of Chicago Press. http://philip.inpa.gov.br/publ_livres/Preprints/2004/FUTURE-book%20chapter.pdf.
6. Renfrow, S. 2006. Mapping the changing forests of Africa. http://nasadaacs.eos.nasa.gov/articles/2006/2006_africa.html.
7. Wallace, S. 2007. Last of the Amazon. *National Geographic.* 211(1):40–71.

Wet Tropical Forests

ONE CANNOT DRAW A PRECISE line de-lineating where one type of forest ends and the other begins. More realistically, the different types of forest blend into one another where they meet, and it some-times requires in-depth knowledge to real-ize that you've moved from one to another. Moreover, the forest types covered in this chapter do not make up an exhaustive list. These are, however, representative of the most typical wet tropical forest biomes.

Mangrove Forests

Perhaps sometimes more correctly identi-fied as swamps than forests, mangrove ecosystems, or **mangals** as they may be called, occupy tropical coastal regions. *Mangrove* is a general term that may in-clude many species of plants distributed among several families, but all are **halo-philic**, that is they are able to grow in a highly **saline** or salty environment, something that most vascular plants cannot. In addition, they grow in the intertidal zone, the part of the shore between high and low tide limits, often in mud that may be **anoxic** or void of oxygen. Mangrove roots are high and stiltlike, spreading out and lifting the plant above the surface of the water, which circulates freely, although slowly, through the maze formed by the roots. Such roots are described as **prop roots**. Without them, it's doubtful that the plants would remain erect in the soft muds in which they grow. The roots are able to survive the low oxygen by producing **pneumatophores**, erect, terminal projections from the roots that stick up into the air above water level for at least part of the day. Air diffuses into the roots through the pneumatophores and provides oxygen.

Mangroves appear to play a number or roles in terms of their overall impact on the environment they occupy. Their roots provide homes for many invertebrate animals like hydroids or oysters that grow attached to them, or, like shrimp, freely swimming among them. Many small fish also live among the roots where they find food and protection from predators, and larger fish may spawn among the roots. Case in point, destruction of mangrove forests in the delta of the Niger River in West Africa resulted in the loss of fish populations, which severely impacted the lives of the people who depended upon them.[1] Above the water line, crabs and other invertebrates may scuttle up and down the trunks and through the branches, as may snakes and lizards. Birds nest among the leaves.

Mangrove swamps are also thought to be important in the formation of land. Sediments carried by rivers to coastal waters settle out as the moving water is slowed by the root network. As sediment accumulates, land forms. Additionally, other debris carried by the moving water, such as driftwood, floating plants, dead animals, and anything else that finds its way into the water can be trapped by the mangrove roots and end up as part of the newly formed land. In Bangladesh, mangroves planted along the shores of the Bay of Bengal entrap sediments that have been eroded off the Himalayas and carried down rivers, generating 300,000 acres (120,000 ha) of new land.[2] Not all of the debris that is captured by the mangrove roots is going to be natural, however. Plastic bags, discarded tires, and trash thrown into rivers, estuaries, bays, or off ships at sea—in fact any kind of flotsam or jetsam—are likely to end up entangled among mangrove roots.

In contrast, the absence of mangroves may encourage coastal erosion. Authors of a study in Viet Nam suggested that coastal erosion occurring in the southern part of that country may be a result of the destruction of mangals in the region.[3] Moreover, Hurricane Wilma in 2005 caused much erosion of the beaches of Cancun in Mexico.[4] The development of Cancun as a resort area has been at the cost of mangrove swamps along the Yucatan Peninsula, and beach erosion has been an ongoing issue there. However, one should not get the impression that mangals, if left to themselves, are invulnerable. In the autumn of 1998, Hurricane Mitch destroyed much of the mangrove swamps of coastal Honduras in Central America.[5]

Finally, mangals are important in maintaining and possibly improving water quality. For example, research in Australia has shown that foul effluents from shrimp farms can be decomposed as they pass through mangrove swamps.[6] Ironically, however, shrimp farming is a major threat to mangrove forests across the tropics. Shrimp farming is often conducted in mangrove swamps where the trees are removed and ponds are constructed. After a few years the ponds are abandoned to avoid disease and reduction in productivity.[2] If new ponds are constructed before the mangrove forests reclaim the abandoned ones, the forests will be destroyed.

Lowland Forest

My home in Upstate New York is on an abandoned farm, in the middle of which are woods that are nothing like a tropical rainforest. In the middle of the woods is a stone fence, totally surrounded by trees and leading one to question how it ever got there and why anyone would build it there. In tropical Belize, one can find Mayan ruins in a similar situation, totally surrounded by trees deep within forests leading to the same questions. However, someone with knowledge of ecology would quickly realize that both the woods on my property and the trees surrounding the ruins in Belize have not always been where they now are. In both cases, the forests that were present originally had been cut down. When the land upon which they had stood was no longer maintained, the forests returned, a process called **ecological succession.** On my property the stone fence once marked the boundary between a pair of adjacent fields; in Belize, the ruins were once a thriving population center. Both have now been reclaimed.

The lowland tropical rainforest is a dynamic ecosystem in which ecological succession occurs constantly. It is the typical rainforest, if such a thing exists; it is the one that is usually illustrated in biology textbooks, encyclopedias, and online. This is where one sees the classic stratification with a nearly barren understory covered by a dense canopy that's punctuated by emergent trees. Indeed, the lowland forest is actually a mosaic of several different types, but all can be roughly divided into two categories: **old growth** or **climax forests**, which are stable, the kind of forest one would expect to find in a given environment, and supposedly unchanging, and **successional forests**, which are recovering from some kind of disturbance and are progressing toward becoming old growth.

Disturbances within a rainforest are not uncommon. Severe rainstorms and runoff, storm surges, sometimes fires, and even volcanic eruption and lava flow, in addition to human activity, can cause widespread damage. Lightening striking or a tree dying and falling over can cause localized damage. In either case, the standing vegetation is destroyed, sunlight is able to penetrate to the forest floor, and either dormant seeds or those randomly carried in from elsewhere germinate. If the disturbance is local or minor, causing gaps or streaks in the standing vegetation, or if a patch of agricultural land is abandoned but plants of some kind cover the ground, the type of succession is usually described as **secondary**; if the disturbance is widespread and major, to the point where the soil is essentially sterilized as may be the case following a volcanic eruption and lava flow or serious fire, succession will still occur, eventually. In this case, it's known as **primary succession**, and the first plants to colonize the area is called a **pioneer community**. These are usually sunloving plants. If the damage is not too extensive, recovery may be complete. However, in cases of really extensive damage, this may not be so. For example, areas of rainforest in Viet Nam that were defoliated in the early 1960s became grassland and had not returned to forest 10 years later.[7] Likewise, if enough rainforest is destroyed so that soils are altered and water cycling is interrupted, complete succession may never occur and an entirely different type of climax, such as a dry savannah or bamboo forest, may develop.

Sometimes points of confusion in the vocabulary about succession may occur. For example, earlier the term **jungle** was used to describe the tangled vegetation along waterways and at other rainforest margins. Indeed, successional rainforest is often likewise thick and tangled, especially when there's no real canopy to block sunlight, and the term *jungle* may be used to describe it. Another is one I encountered in Costa Rica. The naturalist guides at the La Selva Biological Field Station in northeastern Costa Rica, and elsewhere in the country, often use the terms primary and secondary to describe old growth and successional forests respectively rather than to describe the type of succession occurring.

Whatever states of succession in which one happens to find the lowland rainforest, it is still the most common type of rainforest on the planet, and there are several subtypes.

Terra Firme Forest

If the lowland forest is the typical rainforest, the terra firme forest is the typical lowland forest. It's the one that occupies the majority of the Amazon basin as well as coastal regions of most tropical landmasses, and it is the principal subject of this book. The soil is usually not flooded, nor is it particularly fertile, trees support abundant epiphytes, and the stratification of the forest is evident with the canopy punctuated by very tall emergents. In my experiences in Central America, this is where one sees palm trees and huge forest giants with flaring buttress roots most abundantly. The variety of organisms or **biodiversity** is very rich in terms of plants, animals, and probably microorganisms as well. In regions where there is no evident dry season, the forest is probably going to be green all year. In more seasonal forests, there may be individual trees that will drop their leaves and go dormant during the dry periods. Such forests may be described as semievergreen. They compose most of the African rainforest. Plants that do not experience complete leaf loss usually have roots that penetrate deeply into the soil to find water.[8]

Igapo Forest

Igapo forests are perhaps the most dramatic of the tropical rainforests and are most typically found in the Amazon basin. These occur along **blackwater rivers**, such as the Rio Negro and are annually flooded most probably by snowmelt from the Andes to the west (Figure 4-1). Waters may rise as much as 30 feet (10 m) or more and cover miles of forest floor. The roots and lower trunks of the trees are immersed in water; the crowns remain high and dry

Figure 4-1 Flooded forest.

and continue to support epiphytes and arboreal animals. Flooded forests make up only a fraction of the rainforests, and the plants found in them must be adapted to seasonal inundation. Moreover, blackwater rivers are characteristically mineral and sediment poor. Their color comes from the abundant dissolved organic material they carry, but the soil they flood is usually highly infertile. Consequently, igapo forests show much less plant diversity than do terra firme forests.[9]

The flooding delivers water, fish, and other aquatic life to lakes and ponds lying outside the main channel of the river, and such flooding may also help in propagating some of the trees as well. Fruit that ripen during the flooded period fall into the water. Some may be carried downstream to lodge and decay, letting seeds within them germinate. Others may be eaten by fish, which then scatter the indigestible seeds when they defecate. If the seed falls in an area that will be dry when the water recedes, it can sprout. Still other seeds may have to pass through the intestine of a fish in order to have a hard, outer covering softened. In any event, the fish are important in the perpetuation of the igapo species, and the fish seasonally depend on the plants as well. Without the seasonal flooding, both fish and trees would suffer.

Vàrzea Forest

The vàrzea is another kind of flooded forest found in the Amazon basin, and much that was said in the previous section applies to it as well. However, the rivers that inundate the vàrzea are laden with sediment that is deposited on the flooded ground, enriching it. Consequently, the soil is more fertile, and there is greater plant biodiversity in the vàrzea than in the igapo, although less than in the unflooded lowland forest.[8] The rivers of the vàrzea are called **whitewater rivers**, which is a bit misleading. Neither are the rivers violently tumbling over rapids, as is the popular perception of a whitewater river, nor is the water white in color. Rather it's muddy because of the suspended sediment.

Heath Forest

Typically found in Southeast Asia and South America, the heath forest is characterized by dense growth, a low canopy, and an abundance of epiphytes. The soils of the heath forests are sandy and acidic, and insectivorous plants tend to be abundant, perhaps because as insect eaters they are less dependent upon soil nutrients. In addition, plants that have a **symbiotic** relationship with ants also tend to be abundant here.[8] Symbiotic relationships involve two organisms living together, and they may take a number or different forms. They will be described more thoroughly in later chapters.

Peat Forest

Also known as peat swamp forests, these forests are most abundant on the island of Borneo, southwest of the Philippines in the South China Sea. The soils are not seasonally flooded as they are in the vàrzea or igapo forest, but they are waterlogged, perhaps because the rain is unable to drain off, and they are acidic. Fallen leaves are unable to completely decompose; consequently they accumulate and form a dense mat of organic material that is compressed into peat.[8]

Montane Forest

As the name suggests, these are forests that grow in mountains, generally at elevations greater than 3300 feet (1000 m) (Figure 4-2). They are found throughout the tropics but are particularly abundant in Asia, where much of the land is mountainous.[8] The cloud forest described in the first chapter is really a high-altitude montane forest. These are forests in which bryophytes (mosses) are abundant and, in my experience, tree ferns are common. Trees are heavily laden with epiphytes. The soils are less fertile than in lowland forests, and there is less biodiversity than in lowland forests, although it's hardly noticeable to the untrained eye. There are still many different kinds of plants and animals, and some may be unique to a particular territory.[10] The reason for this is the tops of mountains that are higher than 3300 feet are often isolated from one another, more or less as islands are separated from one another by water. Some animals that are adapted to higher elevations are unable

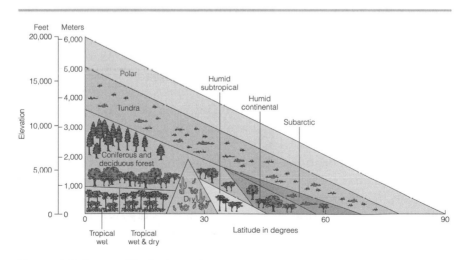

Figure 4-2 Forest altitudinal zonation.

to survive anywhere else; they cannot travel from mountain to mountain. Consequently, these individuals are reproductively isolated from potential mates on another mountain due to geography. In time, gene mutations cause the two populations to become sufficiently different that even if they were somehow brought together, they would no longer be able to interbreed, probably because they would not recognize one another as potential mates. As a result, a single species will have evolved into two separate species, each unique to its own mountain. The same scenario affects plants whose pollen can not travel from one mountain to another. As a result, tropical mountain tops often contain species that are unique.

An example that's often given to illustrate the phenomenon described above is the golden toad (*Bufo periglenes*) that was once found in—and only in—the Monteverde Cloud Forest Preserve in Costa Rica. Measuring perhaps an inch and a half in length, the males were bright orange. The females were a bit larger and were black with red marks edged in yellow. They were found nowhere else. Unfortunately, such unique organisms are extremely vulnerable. Any disruption to their habitat could conceivably cause their extinction, which appears to have been the case with the golden toad.

The montane forest extends to an elevation of perhaps 10,000 feet. At higher elevations, a more typical alpine forest occurs, although the plants one finds may be derived from species found in the rainforest.

Monsoon Forest

Also referred to as tropical deciduous forests, this is an ecosystem that shows obvious rainy and dry seasons, corresponding to winter and summer respectively. Monsoon forests tend to be toward the northern and southern extremes of the tropics, where there is a seasonal change in wind direction. For example, the Asian monsoon that affects India usually begins in June and ends in early autumn. During that time, the wind blows from the ocean carrying moisture. At other times, the wind comes from the northeast, across the Asian continent. Monsoon rains can be very heavy. Indeed, they have been known to cause serious flooding, particularly in Asia.

Biodiversity in the monsoon forest is less than in the lowland tropical forest, but there is still variety. In Nepal, for example, tigers prowl among the plants, which grow more densely than in the true rainforest. Because many monsoon plants drop their leaves during the dry season, more light penetrates to the forest floor, and a denser understory is able to grow as a result.

A Note of Concern

One issue not yet addressed is that of deforestation. Indeed, the subject is discussed in a subsequent chapter, but for the moment it should be mentioned that all rainforest types should be considered threatened, if not at-risk, environments. A number of human activities, including logging, farming, mining, and even the need for firewood, have resulted in deforestation in practically every kind of tropical forest on the planet. This is worrisome for several reasons, perhaps among the most critical is that given the vast diversity of plants and animals in these ecosystems, there are many species that have not yet been discovered. Some of these could prove to be of great value to humans. But if the forests are destroyed, a species could become extinct before its existence is recognized, and any value it may have had will be lost.

In the face of a growing human population, expanding technology, and energy demands, more and more pressure to exploit the rainforests for whatever resources they can provide will be generated. Whether or not they'll be preserved remains to be seen.

References

1. O'Neill, T. 2007. Curse of the black gold: Hope and betrayal in the Niger Delta. *National Geographic interactive edition.* http://www7.nationalgeographic.com/ngm/0702/feature3/index.html.
2. Warne, K. 2007. Forests of the tide. *National Geographic interactive edition.* http://www7.nationalgeographic.com/ngm/0702/feature5/index.html.
3. Mazda, Y. et al. 2002. Coastal erosion due to long-term human impact on mangrove forests. *Wetlands Ecology and Management* 10:1–9.
4. Schwartz, J. 2007, March 4. Shifting sands plague Cancun's restored beach. *Atlanta Journal-Constitution.* http://www.ajc.com/travel/content/travel/beaches/stories/2007/ 03/01/0304springcancun.html.
5. Cahoon, D. R., and P. Hensel. 2002. *Hurricane Mitch: A regional perspective on mangrove damage, recovery, and sustainability.* USGS Open File Report 03-183. http://www.nwrc.usgs.gov/hurricane/mitch/Cahoon%20Regional%20Report%20Final%20Revised.pdf.
6. Trott, L. A. and D.M. Alongi. 2000. The impact of shrimp pond effluent on water quality and phytoplankton biomass in a tropical mangrove estuary. *Marine Pollution Bulletin* 40(11):947–951. http://www.gpa.uq.edu.au/courses/ENVM/7102/2005/Trott%20et%20al.pdf.
7. Westing, A. H. 1971. Ecological effects of military defoliation on the forests of South Vietnam. *BioScience* 21(17):893–898. http://links.jstor.org/sici?sici = 00063568%2819710901%2921%3A17%3C893%3AEEOMDO%3E2.0.CO%3B2H& size = LARGE&origin = JSTOR

8. Whitmore, T. C. 1998. *An introduction to tropical rain forests*, 2nd. ed. Oxford, UK: Oxford University Press.
9. Kricher, J. 1997. *A neotropical companion*. Princeton, NJ: Princeton University Press.
10. Mongabay.com. 2008. Tropical rain forests: types of forests. http://rainforests.mongabay.com/0103.htm.

Plants of the Wet Forests

5

WHEN I WAS AN UNDERGRADUATE, the number of species of organisms on Earth was estimated to be somewhere between 1.5 and 3 million. Today, some biologists have suggested that there may be that many species of organisms living on a single rainforest tree. Nobody knows just how many species of organisms there are, but estimates as high as 100 million can be documented. However many there may be, it's safe to say that probably most have yet to be described, and most of those probably live in the rainforest.

The total number of plant species on the planet is likewise unknown. It probably numbers less than a million, and the greatest variety again is found in the rainforest. Even the number of known species found in the rainforest is too large to cover in a single book, let alone a chapter. Consequently, this discussion is very general. Because the most evident plants in the rainforest are the trees, we'll begin with them, moving then on to the climbing plants and finishing with the epiphytes.

Trees

In general, rainforest trees fall into the two major categories of seed plants: the **angiosperms** or flowering plants and the **gymnosperms** or nonflowering plants. In addition, among the nonseed plants, there are species of **Pteridophytes** or ferns that grow to tree size. These are known as tree ferns, appropriately enough, and we'll begin with them.

Tree Ferns

As the name suggests, tree ferns (Figure 5-1) are ferns with an apparently woody, erect stem that supports the fronds well above the ground. Superficially, they resemble palms. The majority of species are tropical; my own experience suggests that they are most common in more moist environments, such as the cloud forests and the rainy sides of the Hawaiian Islands, and they appear to be abundant on the island of Tasmania off Australia. They are found in tropical forests throughout the world. Tree ferns are living fossils. They predate the seed plants, the dinosaurs, indeed all

Figure 5-1 Tree fern.

terrestrial vertebrates, having dominated terrestrial ecosystems more than 360 million years ago. Many coal deposits are remnants of tree fern forests.

Tree ferns are not true trees in that they do not produce seeds, nor do they have a woody trunk that forms annual rings as the plant grows. Instead, they reproduce by spores, as do the more familiar ferns, and the underground stem or **rhizome** that is found in temperate ferns grows vertically in tree ferns. The importance of this is that it elevates the fronds (leaves) above the ground, possibly relieving them of some competition for light.[1] In gaps or other successional environments, this could be a critical factor in survival. In general, ferns are plants of the understory; consequently, one would infer that they tolerate shade. This is another critical characteristic in the rainforest, where the taller true trees would overshadow the tree ferns. It has been shown that more shade-tolerant plants survive better under such conditions.[2]

Tree ferns are sometimes planted as ornamentals, and some varieties are edible. However, there's not much general information about them. Most interest in them appears to be scholarly.

Gymnosperms

This group of plants is well represented in the northern hemisphere by evergreen trees: the familiar pines, spruce, firs, and the like. However, they do not appear to be well represented in the tropics and are entirely absent from Africa, Australia, and South America (unless introduced) and are represented by only a single species in Central America and by one in Southeast Asia, the Philippines, and the Malay archipelago. Other conifers occurring in Asia are temperate varieties that have extended their range.[3] The so-called Australian pine is not a true pine. The Central American species, the Caribbean pine, ranges from Mexico south to Nicaragua as well as the Caribbean islands.

The gymnosperm trees that are typically found in tropical rainforests are the cycads, although there is another group, the gnetophytes that are represented as well. Some botanists think the cycads may belong in a group of their own. Although they bear cones like the conifers, they also produce swimming sperm cells as do more primitive plants. Though not as old as the tree ferns, the cycads can be considered living fossils, having reached their peak around 150 million years ago during the Jurassic period. That would make them contemporaneous with the dinosaurs. The populations that exist today are relics of formally widely distributed organisms.[4] Moreover, though they're pretty common in rainforest situations, they grow in subtropical areas as well, and one species is known to grow in the mountains of eastern Africa, where it must deal with dry conditions and sometimes snow.[5]

Cycads, like tree ferns, superficially resemble palms. The so-called sago palm that's often planted as an ornamental is actually a cycad. Other cycads have been planted as ornamentals as well. Consequently, cycads in the wild are becoming endangered, not only because their habitat is threatened but also because they are being collected as well.

Another fact about cycads is that some are known to house nitrogen-fixing cyanobacteria (blue-green algae) in specialized roots that grow on the soil surface. These organisms most likely provide the plant with nitrogen and allow it to grow in the poor soils common to rainforests.

The gnetophytes are mostly woody vines, but there is a single species, *Gnetum gnemon* found in the forests of the western Pacific islands and southeast Asia that reaches tree size.

Angiosperms

This is the category of familiar flowering plants, including most of the deciduous trees common to North America. It also comprises the most evident plants of the rainforest, including, of course, the trees, some of which can be absolutely enormous.

Although the rainforest might seem to be an ideal habitat in terms of climate, there are many challenges that plants have had to adapt to in order to be successful there. Perhaps the first is the soils, which generally are nutrient poor. The nutrients that are present are usually just under the litter, such that it is, that covers the soil. To overcome the soil infertility, many rainforest trees have symbiotic relationships with microorganisms that fix atmospheric nitrogen, specifically legumes in Africa and South America and diptocarps in Southeast Asia. In addition, tropical rainforest soils are often shallow, and tree roots do not penetrate deeply into the ground. That would leave many of the taller trees on a very unstable foundation. These plants compensate by developing roots that are modified for the conditions. Exceptionally tall trees, such as kapoks or Central American mahoganies, have buttressed roots that grow outward from the bottom of the trunk (Figure 5-2a). Others have prop roots. These are **adventitious roots** that often grow out from the base of the trunk and encircle it, more or less resembling a grass skirt (Figure 5-2b). In both cases, the modified roots broaden the base of the tree, providing support.

Another challenge that trees may face is **herbivores**, animals that eat plants. Oddly enough, it's often not the large animals that inflict the greatest damage; it's usually insects. Plants have evolved a number of solutions to this problem. Many produce compounds that are toxic to anything that tries to eat them. Others are literally armed with spines. Still others may use what would

Figure 5-2a Buttressed roots of a large tree.

Figure 5-2b Prop roots on a tree.

be described as a morphological defense. Numerous tropical trees form leaves with holes. A suggested reason for this is that butterflies looking for places to lay eggs on foliage that could provide food for the caterpillar once the egg hatches will avoid such leaves because they appear to be already occupied by a leaf-eating caterpillar. Of course, a butterfly would not make that kind of analytical decision; but it is conceivable that instinct would lead a butterfly away from a situation where competition would potentially exist.

Perhaps one of the more effective adaptations by some plants literally amounts to hiring bodyguards. Plants referred to as **myrmecophytes** secrete nectar that attracts ants. Some, such as certain *Acacia* trees even provide hollow stems or thorns in which the ants establish colonies. In many cases then, the ants protect the tree. Some may do so by clipping nearby plants to prevent them from shading the host plant, others may eat caterpillars or aphids that may try to exploit the plant, but still others will physically attack any animal that lands on or even brushes up against the tree.[6] On some trees, one can even see ants patrolling, as I did in Belize. I then watched a guide jostle the tree with a stick and saw the ants swarm out onto the branches and twigs, apparently looking for whatever had disturbed their plant. Parenthetically, the guide was quick to step back from the tree as soon as he bothered the ants, and I followed very quickly.

One additional challenge to plants in the rainforest is the amount of rain they receive. Conceivably, water accumulating on plants could promote mold growth or otherwise harm the plant. The upper surface of leaves are generally waxy, which repels water, and leaves often come to a point, a so-called drip tip to facilitate the removal of water. Rain or condensation water rolls across the leaf surface and over the tip as if it were being poured from a pitcher.

Perhaps the tree that's considered to be most typical of the tropics is the palm (Figure 5-3), or more appropriately the palms. There are many different kinds. These are something of an oddity in that they are more closely related to grass than to any of the more typical woody trees, and they do not reach the heights of the canopy and emergent trees. Moreover, in my experience, one is more likely to find palms, unless they've been deliberately planted, on rainforest edges, in coastal areas, on islands, and along rivers, where they are able to get sunlight. Palms are common in the American and Asian tropics, but not in Africa.[9]

Climbing Plants

Although climbing plants are generally referred to as vines, my experience in Central America has the term being used specifically for soft-stemmed climbers. Woody vines are called *lianas*. These can form dense networks often hanging loosely from host trees, while the soft-stemmed vines cling tightly to their host. Both types of climber basically follow the same pattern,

Figure 5-3 Palm tree.

however; they are generally unable to support themselves, so they grow up a tree in order to reach the canopy and the light that is available in it. Some of these climbing plants may grow from tree to tree once they have climbed into the canopy, providing a highway, so to speak, for arboreal animals.

Some types of climbers appear to get their start whenever a gap in the forest occurs, perhaps by a tree falling after being struck by lightening or from the weight of epiphytes on it. In either case, sunlight reaching the forest floor may encourage the germination of dormant seeds. Once sprouted, the plant grows across the gap until it encounters a tree that it can climb. It then attaches to the bark of its host and grows upward. In a few cases, the roots of the vine may eventually degenerate, and the vine becomes a true epiphyte. Others may actually begin as epiphytes with a seed sprouting in the canopy or on a tree branch and then growing downward, as well as upward into the canopy. When the plant encounters the ground, roots develop. Technically, climbing plants that are rooted and get water and minerals out of the soil may be referred to as hemiepiphytes.[7] True epiphytes are independent of the soil. In general, climbing plants are not parasites of their tree hosts in that they do not depend on them for water or nutrients, only for physical support. There are a few exceptions, however. They include some vines of the morning glory and laurel families.[7] Perhaps the biggest threat to trees by lianas is that too many lianas may weigh a tree down to the point where it can topple over.

Among the odder climbing plants are the stranglers, and the largest single genus of these is *Ficus*, the figs (Figure 5-4). This is a huge genus, containing perhaps 1000 species upon which many types of animals, including apes and monkeys, depend. Different species of figs bear fruit at different times of the year; consequently, figs may be one of the few types of rainforest trees that can be considered **keystone species**. If they were to go extinct, the animals that rely on their fruit for food would be seriously impacted, as would the predators that eat those animals. Figs actually begin life as epiphytes. Their seeds are indigestible; an animal that has eaten a fig passes the seed when it defecates. If the seed lodges on a branch or somewhere in the canopy, it sprouts and sends shoots down the trunk of a tree. Once the shoots reach the ground, they form roots that begin to compete with the host tree for water and mineral nutrients. Several shoots may be suspended from a single tree. These tend to wind around the trunk, intertwining with one another to form what looks like lacework. As the shoots continue to grow, they thicken and fuse together, putting more and more pressure on the tree that is supporting them, eventually blocking the flow of water and nutrients within it. Eventually, the tree dies and decomposes, leaving a hollow fig tree that may reach a height of 150 feet. Mature fig trees may be among the largest trees in the forest, and they are found in rainforests all over the world.[8]

Epiphytes

By definition, anything that grows on a living plant, a fungus for example, can be considered an epiphyte, but the term generally means a green plant that grows on another green plant. Perhaps the best example for North Americans would be the Spanish moss that one can see on trees growing in the deep South. In the tropical rainforest, each tree can seem like its own ecosystem with epiphytes on the trunk and branches. Sometimes, the branches are so thickly encrusted with epiphytes that one cannot clearly see them. Epiphytes are not parasitic; they are able to obtain moisture from condensate on the host tree or from the air by means of aerial roots, and they can often get nutrients from the detritus that accumulates on branches or in the crotches of host trees. In a cloud forest, roots of epiphytes sometimes seem to hang as thickly as macramé. They use the host tree for support only, as by growing on it, in the canopy mostly, and they can take advantage of light that would

Figure 5-4 Strangler fig.

otherwise be unavailable to them. Other than where they grow, they're typical plants with leaves, flowers, fruits, seeds, and sometimes symbiotic relationships of their own, as with some southeast Asian epiphytes and ants.[9]

A vast variety of plants is represented among the epiphytes, including mosses, ferns, and all manner of seed plants, including cactus. Though typically thought of as desert plants, cacti, because of their ability to take advantage of little available water, easily grow as epiphytes in the rainforest. However, the two groups of plants that often come to a botanist's mind when epiphytes are mentioned are the bromeliads, which include the nonepiphytic pineapples, and the orchids.

Most people are acquainted with orchid blossoms and are perhaps aware of the vast horticultural and commercial interest in the plants. However, the orchid family is more than simply a bunch of pretty flowers; it is the largest family of the flowering plants with perhaps as many as 30,000 species, and it is the most diverse. And while the majority of them are tropical, and many of those are epiphytic, orchids are found in almost every terrestrial environment on earth, and they include commercially important plants such as vanilla and possibly medically important ones as well.[10] The tropical orchids are particularly fascinating. Because of their diversity, they are excellent subjects for studies in evolution and coevolution. In the latter instance, they show a number of interrelationships with other plants, animals, and even fungi. Unfortunately, and not simply because of their beauty, some orchid species are endangered because of illegal harvesting and trade. In Africa alone, 85 species are at risk of extinction because they're being collected for food. All orchids are currently protected.[11]

The other plant group that is typically thought of as epiphytic, especially in the American rainforests, is the bromeliads. As mentioned earlier, this group includes the pineapples, which are not epiphytes, but many kinds, including Spanish moss, are. Bromeliad leaves grow in concentric rings, called rosettes, and rainwater and condensate often get trapped between the rings. Consequently, bromeliads are sometimes referred to a tank plants, and arboreal animals will drink from them. More than that, some bromeliads essentially hold micro-ecosystems of their own, with populations of insect larvae, such as mosquitoes, living and growing in the water, feeding on microorganisms also living there. Some species of tree frogs lay their eggs in bromeliads, and the tadpoles feed on the insect larvae. Even crabs may live in the canopy and breed in bromeliads, and some damselfly species breed there as well, their larvae feeding on the crab larvae and even tadpoles. In addition, some birds nest specifically on bromeliads, and some hummingbirds are specific pollinators. Worms, snails, and salamanders also take advantage of the "aquatic" environment bromeliads have to offer. Perhaps bromeliads too should be considered keystone species, as many kinds of organisms would be displaced were anything to happen to them.

Any biology student who has taken a parasitology course has probably learned the poem:

Big fleas have little fleas,
Upon their backs to bite 'em,
And little fleas have lesser fleas,
and so, ad infinitum.

Perhaps if I were more creative, I could come up with a similar ditty for rainforest trees, for the same rule applies to them and their epiphytes. Large bromeliads may support mosses, lichens, liverworts, or algae. Even temperate forest canopies offer a more complex world than was originally thought, and whenever a tree falls, especially in a tropical rainforest, everything that was supported by its branches falls with it and dies. Many of these arboreal organisms have not yet been described, and some may hold chemicals that are potential medications. More than one biologist has argued that the cure to cancer may very well be growing in a tree in Africa or South America. It could just as easily be growing in Australia or Sumatra, or on an island in the Pacific. The point is that the epiphytes of the tropical rainforests have much to offer us, if not in commercial products, then in knowledge. If the rainforests are destroyed, then they too will be destroyed, along with all potential resources they may have had.

Additional Considerations

The forgoing discussions only begin to address the diversity of tropical rainforest plants. For one thing, orchids and bromeliads are only two of the plant families that make up the epiphytes. Likewise, epiphytes are not restricted to the canopy. Some, particularly mosses, lichens, and ferns, grow on tree trunks. And despite the scarcity of light in the understory, there are plants adapted to it as well. Young trees, lianas, and members of the genus *Heliconia* (Figure 5-5) are examples. Neither does the discussion address the fungi and bacteria that live in the soils. Though not plants, their existence can be vital to the survival of the plants.

Perhaps this is a good point to mention the introduction of alien species into the rainforests, because the practice can be destructive. Many of the alien plants that have been introduced initially were done so for food. For example, the sugarcane grown extensively in Brazil and Central America was originally domesticated in Papua New Guinea. Bananas, also grown in Central America, originated in Southeast Asia. Today, one can see vast plantations of these plants growing where rainforest once stood. Such plantations disrupt not only the forest that once occupied the land on which they are located, but also the rivers that drain those lands and the estuaries and salt water into which the rivers flow, as fertilizer and pesticides run off the culti-vated fields and debris is carelessly discarded. More-over, the disruption can affect people. One Costa Rican her-petologist I know argues that the *fer-de-lance*, a poisonous snake, is increasing its pop-ulation in Central America because of abandoned maca-damia nut plantations. The abandoned plantations at-tract rats, which feed on the fallen nuts. The snakes, in turn, are attracted by the rats, which they eat. The animals that control the snakes have not yet followed them onto the plantations; consequently, with abundant food and no enemies, the snakes prosper, and more frequent snakebites in Costa Rica is the result.

Figure 5-5 Heliconia (bird of paradise) plant.

Part of the problem with large plantations is they greatly simplify an ecosystem, and it has been argued that the integrity of an ecosystem is pro-portional to its complexity. Simple agricultural ecosystems, often called mono-cultures, can be vulnerable because of this simplicity. Consequently, much energy in terms of activity and chemicals are often needed to keep the mono-culture functioning. A banana plantation, for example, is seen as a dinner call to the pest organisms that exist at the bananas' expense. If these pests are not controlled, the bananas may become valueless. Who, after all, would buy a rot-ted or insect-eaten banana? Consequently, pesticides are necessary. And rain-forest soils are nutrient poor; therefore, fertilizers are needed. Of course, it has been argued that the financial benefits of such plantations justify their existence by their financial contributions. That argument has been taken up by Vandermeer et al.,[12] and I would encourage anyone who is interested to pur-sue it there. However, I would mention that the financial argument has been used to justify many ill-conceived projects. A case in point is the destruction of rainforests to build Australian eucalyptus plantations. It was argued that these trees would provide an abundant lumber crop. Unfortunately, eucalyptus was not well adapted to many of the places it was planted, and the projects failed.[13] Even where eucalyptus has grown successfully, it has had its consequences. I've seen pure stands of eucalyptus in Costa Rica where the trees stand totally void of vines, lianas, and epiphytes that are normal to that part of the world.

Even plants that have been imported innocently, for ornamental purposes or backyard gardens, can escape. For example, I've seen *Coleus*, a native of Africa, growing wild in Central America, where it was crowding out native plants. Alien plants can disrupt the ecosystems they invade, causing the loss of more desirable, native species. In short, wherever possible, ecosystems are best left in their natural state.

References

1. Barclay, I. 2002. *Cold-hardy tree ferns*. http://www.angelfire.com/bc/eucalyptus/treeferns/.
2. Durand, L. Z., and G. Goldstein. 2004. Photosynthesis, photoinhibition, and nitrogen use efficiency in native and invasive tree ferns in Hawaii. *Oecologia* 126(3): 345–354. http://www.springerlink.com/content/hm05n415tjfphd4f/.
3. U Ang Din. 1958. Pines for tropical areas. *Unasylva* 12(3). http://www.fao.org/docrep/x5388e/x5388e03.htm#distribution%20of%20tropical%20pines.
4. Anon. 2008. *Cycads*. http://www.nd.edu/~fboze/cycads.htm.
5. Speer, B. R., and N.C. Arens. 1995. *Introduction to the cycads: Legacy of the Mesozoic*. University of California Museum of Paleontology. http://www.ucmp.berkeley.edu/seedplants/cycadophyta/cycads.html.
6. Kricher, J. 1997. *A Neotropical companion*. Princeton, NJ: Princeton University Press.
7. University of California, Los Angeles. 2008. *Life forms*. http://www.botgard.ucla.edu/html/botanytextbooks/lifeforms/index.html
8. Benders-Hyde, E. 2002. *Strangler figs*. http://www.blueplanetbiomes.org/strangler_figs.htm.
9. Anon. 2008. *Epiphytes—Adaptations to an aerial habitat*. http://www.kew.org/ksheets/epiphytes.html.
10. Anon. 2008. *Orchidaceae*. http://www.kew.org/scihort/orchids/whatareorchids.html.
11. Graham, S. 2001, August 2. Illegal trade threatens African orchids. *Scientific American.com*. http://www.sciam.com/article.cfm?articleID = 000F38FF-489B-1C60-B882809EC588ED9F.
12. Vandermeer, J., I. Perfecto, and V. Shiva. 2005. *Breakfast of biodiversity: The political ecology of rain forest destruction*, 2nd ed. (revised). Oakland, CA: Food First Books.
13. González, G. 2003. No forest for the trees. *Inside Costa Rica*, June 4, 2003. http://insidecostarica.com/specialreports/environment_no_forest.htm.

Vertebrate Animals I: Mammals

FOR SEVERAL YEARS, I TAUGHT a rainforest course in Costa Rica that included a visit to the La Selva Biological Station of the Organization for Tropical Studies, an international consortium of universities, including the Universidad de Costa Rica, and research institutions that conduct rainforest research. One year, the guide that was going to conduct us through their forest began by asking the students if there were any particular animals they wanted to see. One young man who was a budding herpetologist wanted to see poisonous snakes, particularly a *fer-de-lance*. Practically everyone else wanted to see a jaguar.

If blindfolded and dropped randomly in a rainforest somewhere, a good botanist could probably tell more or less where he or she was by examining a few plants. Most of the rest of us would most likely be clueless, unless we happened to see an animal that we recognize, and most likely, that animal would have to be something big. Indeed, most ecotourists, other than birders, are interested only in the attractive, big animals, what among professionals have come to be known as *charismatic megafauna*, and that interest is completely understandable, because they are the least common, and it is usually the big animals that make a particular environment most evidently unique. Consequently, our discussion of animals will begin with and concentrate on charismatic megafauna. I must express one caveat, however. There is no way that all of the animals in the rainforests can be covered in a book of this nature any more than all of the plants could have been covered in the previous chapter. There are simply too many, both in terms of number and variety. Consequently, this and the two subsequent chapters consider only a representative sampling. It is hoped it will lead to further study by those reading this book.

The Great Apes

This group includes four Old World animals: three of them, the chimpanzee, the bonobo, and the gorilla are African, and the orangutan is Asian. Apes, technically, are not monkeys. In fact, they are more closely related to humans than to true monkeys. Until fairly recently, misinformation on these animals was legion. For example, it was popularly believed that the gorilla was a violent, aggressive animal, while the chimp, in contrast, was peaceful and comparatively civilized, even cuddly. However, the work of Diane Fossey[1] has shown clearly that the gorilla is the more peaceful animal; it prefers to be left to itself. On the other hand, Jane Goodall[2] found that the chimp could be extremely violent, even murderous. In addition, it was believed even among scientists that all of the apes are members of the same family, Pongidae, and were more closely related to one another than any of them was to humans, family Hominidae. More recently, DNA studies have shown that this is not the case. Chimps and humans are more closely related to each other than either is to any of the other apes. Indeed, it has been argued that chimps and humans belong in the same genus.[3]

Gorillas are distributed unevenly across central tropical Africa where they inhabit a number of forest habitats, from lowland to montane. There are two species, *Gorilla gorilla* and *G. beringei*, the latter restricted in range largely to a pocket in eastern Zaire. Each species is divided into at least two subspecies. All gorillas are vegetarian, eating leaves, stems, and shoots of plants. Eastern populations of *G. gorilla* also include fruit and insects in their diets. In general, gorillas live in groups that are dominated by an adult male **silverback**, along with adult females and their young. Subordinate adult males may or may not be present, and the number of animals in the group varies among the different subspecies, but it may range from as few as two to more than 20. It is the silverback that keeps the group intact; he protects the juveniles, who are most likely his offspring, and he mates with the females. Should he die, the group may disband and its members disperse, it may fragment, or it may remain intact if one of the young males is able to take over.[4]

As mentioned above, the chimpanzee (*Pan troglodytes*) (Figure 6-1), along with the bonobo (*Pan paniscus*), is the closest living relative to human beings. Its DNA may be more than 99% identical with ours. Humans and chimps diverged from a common ancestor perhaps six million years ago. In contrast, the line that led to the gorillas split off the chimp/human line much earlier. Additionally, chimp behavior is more like ours than is the gorilla's. Chimps live in but are not restricted to the rainforests of central and western Africa in large troops with as many as 75 members, though usually much fewer. The troop is dominated by a group of related males, one of which is usually the leader, but all males can mate with the adult females in the troop. The leaders, however, often maintain the integrity of the troop with brute force and violence. Their diet is based around fruit, but they will also eat birds eggs and insects, particularly termites, which they have been observed to "fish" for by poking sticks into their nests and then eating the termites that grabbed the stick.[2] This was the first observed instance of tool use by a nonhuman primate. More instances of tool use have since been observed, forcing us to

Figure 6-1 Chimpanzee.

change our definition of what constitutes being human. Moreover, chimps have been observed to hunt cooperatively for, kill, and eat other mammals, including other primates, behavior common enough in humans but not observed in gorillas. In addition, chimps have also been seen killing one another, behavior that's much too common in humans but rare in gorillas. And if that were not enough, chimpanzees have been known to practice cannibalism, usually an adult eating an infant.[5,6] Consequently, regardless of how cute, intelligent, and human chimps may appear to be, past attempts of trying to impose human characteristics on the animals had resulted in a completely inaccurate picture of the animals' behavior. They are simply products of their evolution, and evolution is random. It favors whatever works.

The bonobo was once known as the pigmy chimpanzee. It looks very much like the chimp, although it is slightly smaller and more slender, and its hair color and pattern also differ slightly from that of the chimp. The two species are very closely related, having diverged from a common ancestor only two to three million years ago. Bonobos occupy successional swamp forest between the Congo and Kasai Rivers in western Zaire, a restricted area, and their diets are largely based on fruit, although they eat other vegetation and some simple animal material.[7] Apparently they do not hunt, as do chimpanzees, and they seem to have a more gentle disposition. Strong female-to-female bonds appear to have raised the females to higher levels of dominance within bonobo troops than among chimps, and sexual activity is apparently very important in maintaining troop integrity.[8] Those interested in bonobos can find more in *The Forgotten Ape* by Frans B. M. de Waal.[9]

The last of the great apes, and the one least closely related to man, is the orangutan. Again, there are two species: one on the island of Sumatra (*Pongo abelii*) and one on Borneo (*Pongo pygmaeus*). Incorrectly believed to be named for the orange tint to its hair, it is restricted to the lowland diptocarp rainforests of the islands. It is the only Asian ape. Moreover, it is the most arboreal and solitary of the great apes, spending practically all of its time in the canopy and rarely coming to the ground. Fruit is the major component of the orangutan's diet, but it also eats other vegetable material, and the species prefer the older forests to the earlier, successional forests. On Sumatra, the orangs rarely come down from the trees; on the ground they're vulnerable to predation by tigers and clouded leopards. Those large cats are not a threat in Borneo, and the orangs there are seen more frequently on the ground.[10]

One unfortunate characteristic that all apes share is that they are endangered. Gorillas and chimpanzees in Africa are hunted for what is referred to as bush meat, wild game that is sold often on the black market, or animals are illegally captured to be sold as pets or to zoos. In addition, an increasing human population puts the animals at risk because of habitat loss, as people clear more and more forest for agriculture, firewood, and living space. Moreover, as humans and apes are forced into closer contact, the threat of disease being passed to the apes exacerbates their dangers. As will be discussed, apes are capable of self-medication to a limited extent, but they do not have the means of treatment that we do. In Asia, the orangutans and gibbons are similarly threatened. The loss of these animals would be tragic enough esthetically, but there is no way of knowing if it might turn out to be tragic to us biologically. The great apes are our closest living relatives as a species; we may have much to learn about ourselves by studying them. That alone should make them valuable to us.

Lesser Apes and Other Primates

The lesser apes fall into a number of species in a single genus and are collectively known as gibbons. Their line diverged from that of the great apes and humans roughly 20 million years ago. The gibbons are much smaller than the great apes and even more arboreal, rarely coming to the ground. In the canopy, they are extremely agile, gracefully swinging from branch to branch, or leaping from tree to tree in old growth tropical rainforests. Gibbons live in southeast Asia. They are opportunistic feeders with fruit making up the majority of their diet. They live in family groups of a mated-for-life pair of adults and their juvenile offspring, which disperse when they reach maturity.[11]

Most remaining primates are classified as monkeys, a very diverse group that includes many members that are not rainforest inhabitants. In general, there are two subgroups: the New World monkeys and the Old World monkeys. The two diverged perhaps 30 million years ago; they differ in terms of where they live, as their names suggest, and anatomy. The New World monkeys tend to have flatter noses and prehensile tails. As a group, they are more arboreal than the Old World monkeys, which include many ground-dwelling species. One particular genus of New World monkey an ecotourist is likely to encounter, though not necessarily see, is the howlers (*Alouatta spp.*). These are strongly arboreal; the males have enlarged larynxes and throat sacs that serve as resonating chambers. Consequently, the animals are capable of loud, impressive vocalizations. The animals typically howl early and late in the day as a means of marking territory. The first time I heard one, I thought I was listening to a jaguar. Kricher gives a good summary of the howlers and other New World monkeys.[12]

One particularly interesting group of primates is the lemurs. Lemurs are found only on the island of Madagascar and the nearby Comoro Islands off the east coast of Africa. These are primitive, mostly arboreal primates thought to resemble the ancestors of monkeys and apes. They, along with other primitive primates, are collectively known as **prosimians**. At one time, nearly all of Madagascar was tropical forest, of which roughly 20% remains. The rainforest there is under similar stresses to that in Africa, and the same fate probably awaits it. Once it is gone, the lemurs will be extinct, at least in their native habitat.

The remaining prosimians are small, nocturnal, arboreal animals. They are found in Africa and Southeast Asia, principally, if not entirely, in rainforests. They are normally not seen unless someone who's knowledgeable happens to be looking for them specifically.

Jungle Cats

Mythology and popular culture have it that the king of the jungle is the lion. In fact, the lion is an animal of the savannas, at least in Africa. The remnant population of the Asian lion lives in deciduous forest. The big cats of the Old World rainforests are the leopard in Africa and the tiger in Asia, although the leopard occupies the Asian rainforests as well. The leopard is more of an arboreal hunter where the ranges of the two overlap, while the tiger hunts on the ground. Therefore, there would not appear to be much competition between the two. Both of these animals are capable of exploiting different types of habitats; neither is restricted to the rainforest. The black panther, which is thought by many to be another jungle cat, is in reality a leopard. Many kinds of vertebrates have a genetic variation called **melanism**, where an individual is black rather than the normal color. A black panther is a melanistic leopard.

In the Western Hemisphere, the biggest predator in the rainforest is the jaguar, which in coloration looks a lot like the leopard. It too can show melanism. It originally occupied a variety of habitats, but it has largely been exterminated outside of the rainforest.

In addition to the large animals mentioned above, the cat family includes many small to medium-sized animals, and a number of these occur in rainforests. They include the ocelot and margay in the Western Hemisphere, the marbled cat and leopard cat in Asia, and the golden cat in Africa. The American mountain lion or cougar can also be found in rainforests, although it has a very wide range and is seemingly capable of living wherever it is given the opportunity. Like the apes, many of the wild cats are also endangered, and their ranges are being reduced. They are hunted for their hides and in some cases for their supposed medicinal value. They are captured for disreputable zoos, and they are losing habitat to humans. As with the apes, the extinction of any cat species would be a major biological loss.

Bats

While hardly charismatic megafauna, bats are abundant and important members of rainforest fauna, as well as being common virtually all over the world. Within the rainforest, however, one finds not only insect-eating bats as in North America and Europe, one also finds bats that eat fish, frogs, and small lizards. More importantly, however, there are fruit-eating bats that help to disperse seeds, and nectar feeding bats that are vital to the pollination of the plants on which they feed. The fruit-eating bats clearly got their name from their diet. They are a widespread, diverse group of animals, found throughout the tropics and on many islands as well. In the process of eating fruit, they swallow seeds, which pass undigested through their intestines. Before the bat releases the seed, however, it may have flown a significant distance from the tree that produced the seed. Consequently, seeds can be dispersed over a large area by these bats, and this may be important in terms of reestablishing disturbed or fragmented forests. A study by Garcia et al.[13] describes this phenomenon.

Nectar-feeding bats (Figure 6-2) eat pollen as well as slurping nectar and get pollen on their faces in the process, which they pass to subsequent flowers they visit. If it happens to be a **conspecific** of the first, pollination is accomplished. Nectar-feeding bats are active at night; consequently, the flowers

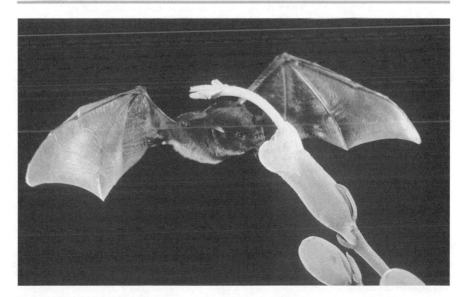

Figure 6-2 Nectar-feeding bat.

they require must also be open at night. Moreover, since bats rely on their sense of smell, these flowers would have to be pungent and fruity rather than brightly colored. The nectar is usually abundant and high in sugar, and the pollen rich in protein.[14]

Most of the remaining bats eat insects, and their existence probably contributes much to our comfort. One group that is an exception and seems to provoke a bizarre fascination, particularly among students, is the vampire bats. These are found in the Western Hemisphere, from the southwestern United States, through Mexico, Central America, and South America as far south as Bolivia and perhaps northern Chile. They inhabit rainforests as well as other environments. Their principal prey is large mammals, and so ranching, it would seem, has worked to their advantage. They do not as a rule attack humans; however, they have been known to feed on human blood. The vampire bat feeds by slitting its prey's skin and lapping up the blood. An anticoagulant in its saliva prevents the blood from clotting, and the bat's teeth are so sharp, the prey probably feels nothing, at least at first. The danger to an animal being fed upon by a vampire bat is not that the bat will bleed him dry. Rather, it is that the vampire bat can spread the viral disease rabies.[12]

Other Rainforest Mammals

Although cats and primates may be among the more interesting of the rainforest mammals and bats may be among the more important, volumes could be written on the remainder. Species of some of our common Northern Hemisphere animals, such as bear and deer, manage to exist in rainforests, in particular in tropical America and in tropical Asia. Some animals, however, have evolved into completely unique species on these continents, and many exploit niches within the rainforests. I find that the best examples of these are in South America.

Other than Australia, South America has been the most isolated generally inhabitable continent throughout geologic time. I do not consider Antarctica to be inhabitable, although numerous penguins and seals would disagree with me. South America's most recent connection, prior to North America, was Antarctica, from which it separated sometime after the break-up of Gondwanaland. It originally held populations of marsupials that were as diverse as those of Australia, if not more so, along with the ancestors of modern anteaters, armadillos, and sloths. However, perhaps 30 million years ago, monkeys and rodents arrived in South America by means that are not clear. Africa and South America were closer then than they are today; the animals could have conceivably made the trip by rafting or island hopping or some combination of the two, from one continent to the other.[15] Once in

South America, the rodents diversified more than on any other continent, as they found a variety of vacant niches they could exploit. For example, the world's largest rodent, the capybara, which can weigh more than 130 pounds (60 kg) lives a semiaquatic life in much of tropical and subtropical South America east of the Andes. Other rodents include the nutria, chinchillas, pacas, agoutis, guinea pigs, and porcupines. Others existed in the past, but periodic connections between North and South America have allowed ancestors of animals like the jaguar and spectacled bear into South America, which may have led to the extinction of some, as most likely did the arrival of humans.

Unique mammals exist elsewhere. For example, Borneo has the pygmy rhinoceros, Australia has its rainforest marsupials, and Africa has rainforest antelopes, but these are evident variations of animals found elsewhere on their respective continents. South America's zoology retains some measure of its uniqueness because it's still somewhat isolated. The Isthmus of Panama is described by some ecologists as a **filter route**: it selectively inhibits some organisms from passing through while allowing others to. For example, the giant ground sloth was able to pass from the south through the tropical isthmus to the north while the capybara was not, although the ground sloth was not able to survive in its new home. Similarly, the cougar was able to migrate in the opposite direction while the bison was not. The isthmus has existed for perhaps three million years, formed by the west edge of the Caribbean Plate being forced upward by the eastern edge of the Cocos plate. Prior to that, connection between the two continents was probably a land bridge that existed along what is now the Antilles Archipelago, the chain of islands that extend from Venezuela through Puerto Rico and Cuba, and eventually to Florida. This is perhaps how tropical plants like the mahogany and strangler fig got to Florida. During periods of low sea level, say during an ice age when a great deal of sea water was locked up in glaciers, the islands may have been connected, or the water dividing them was narrow enough to be crossed easily. The small islands of the Lesser Antilles, the southern part of the archipelago, would have provided a narrow bridge, and so this too would have acted as something of a filter route. Even so, the bridge was temporary, and animals that made the southward crossing far enough back in geologic time still had long enough periods of isolation to evolve into species different from their ancestors in North America. An example of this would be the giant otter found in the Brazilian rainforests. Accordingly, while similarities between Northern and Southern American mammalian faunas exist, there are some clear differences as well, as perhaps best shown, again, by the rodents.

Any description of unique tropical mammals would be incomplete without mention of freshwater dolphins. Actually freshwater dolphins exist in the

Ganges and Indus Rivers in India, which are more monsoon regions than tropical rainforest areas, and the gray dolphin of the Amazon and other South American rivers can live in salt water as well. But the pink dolphin or boto (*Inia geoffrensis*), the largest of the bunch, is truly tropical. It is found in the Amazon and Orinoco basins in South America. It feeds on freshwater fish, including *piranhas*, and crustaceans, and it can sometimes be seen among inundated trees in flooded forests. Unfortunately, like so many other animals, the freshwater dolphins are endangered.

Exotics

As with plants, mammals have been transported around the globe either deliberately or accidentally by humans, and in many cases, they have caused problems. Small rodents, rats and mice in particular, are notorious camp followers. Everywhere humans seem to have gone, these animals have followed, and they have sometimes been destructive. When no predators are present, these animals reproduce rapidly and can overwhelm native rodent species and get into human food supplies. House cats that have been brought with human immigrants to control these pests have often gone feral and damaged native bird populations. The Indian mongoose has been deliberately released on many tropical islands in an attempt to control introduced rats, and while it has successfully gone after the pests, it has also gone after native animals and decimated many. For example, it is thought to be a factor in the decline of the Nene, the native Hawaiian goose.[16]

References

1. Fossey, D. 1983. *Gorillas in the mist*. Boston: Houghton Mifflin Co.
2. Goodall, J. 2001. *In the shadow of man* (Revised ed.). Boston: Mariner Books.
3. Pickrell, J. 2003, May 20. Chimps belong on human branch of family tree, study says. *National Geographic News*. http://news.nationalgeographic.com/news/2003/05/0520_030520_chimpanzees.html.
4. Cawthorn Lang, K. A. 2005. Primate factsheets: Gorilla (*Gorilla*) taxonomy, morphology, & ecology. http://pin.primate.wisc.edu/factsheets/entry/gorilla.
5. Cawthorn Lang, K. A. 2006. Primate factsheets: Chimpanzee (*Pan troglodytes*), taxonomy, morphology, & ecology. http://pin.primate.wisc.edu/factsheets/entry/chimpanzee.
6. Anon. 2006. Chimpanzee: *Pan troglodytes*. http://www.geocities.com/RainForest/Canopy/3220/chimpanzee.html.
7. Cawthon Lang K. A. June 7, 2005. Primate factsheets: Bonobo (*Pan paniscus*) taxonomy, morphology, & ecology. http://pin.primate.wisc.edu/factsheets/entry/bonobo.

8. de Waal, F. B. M. 1995. Bonobo sex and society. *Scientific American,* March. http://songweaver.com/info/bonobos.html.

9. de Waal, F. B. M. 1997. *Bonobo: The forgotten ape.* Berkeley: University of California Press.

10. Cawthon Lang, K. A. 2005, June 13. Primate Factsheets: Orangutan (*Pongo*) taxonomy, morphology, & ecology. http://pin.primate.wisc.edu/factsheets/entry/orangutan.

11. Anon. 1999–2007. *All about gibbons.* http://www.enchantedlearning.com/subjects/apes/gibbon/.

12. Kricher, J. 1997. *A neotropical companion,* 2nd ed. Princeton, NJ: Princeton University Press.

13. Garcia, Q. S., J. L. P. Rezende[1], and L. M. S. Aguiar. 2006. Seed dispersal by bats in a disturbed area of southeastern Brazil. http://www.ots.ac.cr/tropiweb/read/revistas/48-1/ecogarci.htm.

14. Gibson, A. C. 2001. Mildred E. Mathias Botanical Garden Newsletter, 4(4). http://www.botgard.ucla.edu/html/membgnewsletter/Volume4number4/Batsandtheirflowers.html.

15. Texas A&M University. 2001. Rafting rodents from Africa may have been ancestors of South American species. *Science Daily* October 17. http://www.sciencedaily.com/releases/2001/10/011012073342.htm.

16. Tobin, M. E. 1994. Prevention and control of wildlife damage. *Internet Center for Wildlife Damage Management.* http://icwdm.org/handbook/rodents/PolynesIan Rats.asp.

Vertebrate Animals II: Birds, Reptiles, Amphibians, and Fish

7

FOR YOUNG UNDERGRADUATES AND RUN-of-the-mill tourists that might visit a rainforest with the same attitude they'd visit a zoo, the big mammals are the most attractive things one could hope to see. I once crossed paths in Costa Rica with a group of seniors from a cruise ship. They had just visited a commercial reserve that featured a canopy ride, and I heard one of them complain that for what they had paid for the ride, the managers of the reserve ought to have brought out animals for them. I resisted the urge to strangle her. For many serious ecotourists, the birds are the most attractive animals in the tropics, and for others it's the snakes. Moreover, I've seen more than one casual student take a sudden interest when looking at a brightly colored poison dart frog. While the big mammals may have the major charisma, the less conspicuous, more "primitive" animals are every bit as much products of evolution and occupiers of specific **niches**, and many of them would have a significant impact on other organisms if they were to go extinct. We'll consider some of them in this chapter.

Birds

My introduction to rainforest birds began bright and early on my first visit. A group of chachalacas, turkeylike chicken-sized birds, not far from the window of the cabin in which I was staying, greeted the sun with a raucous cacophony. Later that morning I saw my first wild keel-billed toucan and ocellated turkey, and later on the trip I saw my first wild parrot. Subsequent trips have revealed many more species, yet I've hardly scratched the surface. As with most other forms of life, the diversity of birds is far greater in the tropics than in temperate areas, and the diversity is enhanced by what we think of as temperate birds that migrate to the tropics for the winter. In reality, they are probably tropical birds that migrate to the temperate regions for the summer. That would include, for example, four species of tanagers that are found in the western United States.[1] By spending summers in the north, they are able to reproduce with less competition than they would find in, say, Panama, where there are an additional 40 species of tanagers.[2]

South America has been called the bird continent. Its long isolation and relative scarcity of mammalian competitors allowed for the evolution of a large aviary, including one that stood as tall as an elephant, weighed as much as a horse, and ate other animals.[3] The so-called terror bird was unable to compete with predatory mammals once South America connected with North America, but a diverse assemblage of more typical birds remains, as does a collection of unique varieties. Indeed, all of the tropics show large bird diversity because there are so many niches for the birds to fill. Case in point, the northeast United States has only one species of bird, that pollinates flowers, the ruby throated hummingbird. There are not that many species of flowers that produce as much nectar as a hummingbird requires. In contrast, in the American tropics, there are many kinds of flowers that produce abundant nectar, and there are multiple species of hummingbirds as well (Figure 7-1a). Hummingbirds are unique to the Americas, but there are other small birds in the Old World and elsewhere that are **ecological equivalents** of hummingbirds. They fill the same niche and have evolved similar characteristics, a phenomenon sometimes described as **convergent evolution**, or more simply as convergence. These include the sunbirds of tropical Africa and Southeast Asia, the honeyeaters of Australasia, and even species of honeycreepers in Hawaii (Figure 7-1b).

Figure 7-1a Hawaiian nectar-feeding honeycreeper.
Figure 7-1b Tropical American hummingbird.

There is probably no single type of bird that characterizes the tropics as, say, penguins characterize Antarctica. But there are several groups that most people associate with rainforests, correctly or incorrectly, and one of these would be the parrots (Figure 7-2). Parrots are cosmopolitan, although they probably originated somewhere in the Old World. They can be found in all continental tropical forests and on South Pacific islands as well, and a number have adapted to environments outside of the forests. Many kinds have long been kept as pets, and feral populations now exist in the United States. Parrots vary greatly in size and color; many show bright reds or greens or striking patterns, and many kinds appear to take to humans easily and accept domestication. Moreover, they are reputed to be extremely intelligent birds and good mimics, all of which may explain, in part, their popularity as pets. Unfortunately, that popularity has lead to the decline of wild populations of some species, particularly

Figure 7-2 Representative parrot.

in Southeast Asia and Amazonia, as illegal hunters have captured and smuggled wild birds for pets, zoos, and aviaries. Hopefully, captive breeding is now replacing the capture of wild specimens, and pressure on endangered populations is being reduced.[4]

Another cosmopolitan bird group, perhaps more so than the parrots, is the kingfishers. With a representative in temperate America and several in tropical America, the group is much more diverse in the Old World, with many species found in tropical forests. One in particular, the kookaburra, is perhaps as representative to Australia as are kangaroos. In actuality, it is the laughing kookaburra with its raucous call that typifies the bird, but there are other species, and some are found on Papua New Guinea and the Aru Islands of Indonesia as well. These animals feed on a variety of things, including insects, snakes, and small birds; they're not restricted to fish.[5] The diversity of kingfishers is the result of a few ancestral species spreading into new habitats and finding niches they could exploit. With the passage of time, subgroups became reproductively isolated from one another and diverged into separate species, what is referred to as **divergent evolution** or **speciation**. The great biodiversity of the tropics is a result of many opportunities ancestral species found.

If there is a bird that represents the New World tropics, it would probably be the toucan, or, more properly, toucans, as there is more than one species, one of which graces boxes of Froot Loops cereal. That is appropriate as toucans are largely fruit eaters, although they will take insects, small snakes, and even baby birds on occasion. They range from Mexico, through Central America, and into Brazil, and they include smaller birds with less impressive beaks than the typical toucan, although, it is the massive beak that makes toucans unmistakable (Figure 7-3).[6] Superficially similar to, and to some extent ecologically equivalent of, toucans in rainforests of Southeast Asia are the hornbills (Figure 7-4). They too have large, impressive bills, but they are more ecologically and nutritionally diverse, and they are unrelated to toucans.[7]

There are some tropical birds that are popularly believed to be associated with rainforests but are not

Figure 7-3 Toucan.

Figure 7 4 Hornbill.

necessarily. In particular among these would be some of the wading birds, such as the flamingo. Typically pictured to be wading along the edge of a pool with dense jungle behind it, flamingoes have become iconic. In reality, they and other water birds are more opportunistic; they take advantage of water where they can find it. Some, such as egrets and herons, can be equally as comfortable along a farm pond as they are by a rainforest river. Ironically, rainforest destruction has actually opened parts of the tropics for some bird species while destroying habitat for others. Not a wader, the snail kite, for example, is adapted to the shallow water that waders favor, and it has grown in number in Belize as agriculture has expanded at the cost of rainforest. Irrigation ditches and the like have turned out to be suitable habitat for the kite's favorite prey, and the birds have responded predictably. In another case, rainforest destruction for cattle ranching has opened tropical America for an immigrant from Africa: the cattle egret. In Africa, this particular bird is found on the savanna, where it closely follows large grazing animals eating the insects they stir up. In the Americas, it has adapted to following cattle in the same manner.

The final group of birds to be considered here is the raptors, the so-called birds of prey, although some are carrion feeders. Roughly two thirds of all raptors spend part or all of their lives in the tropics, and many are threatened because of habitat loss, chemical contamination, or overhunting.[8] The largest raptor of the Americas is the harpy eagle, a predator of the lowland forest canopy, although it will hunt in open areas adjacent to the forest. It preys principally on arboreal mammals such as sloths and monkeys.[9] A second big rainforest raptor is the so-called monkey-eating eagle of the Philippines. It nests high in the canopy or on ledges, and its diet includes virtually any animal it can get its talons on, including monkeys. It is one of the largest eagles in the world, and one of the most threatened. The bird is restricted to four islands in the Philippines, and rainforest destruction is being carried on there at an alarming rate.[10] Africa's biggest raptor, the martial eagle, is not a bird of the rainforest. The African fish eagle, another large raptor of that continent, lives south of the Sahara Desert where ever there happens to be open water, including in or near the rainforest. Its close relative, the Madagascar fish eagle, is a bird of deciduous forests, and it is endangered because of habitat destruction.

Reptiles

If most tourists to the rainforest want to see charismatic megafauna, some are content with cute and cuddly microfauna. Reptiles do not fit that description. In fact, most people perceive reptiles, snakes in particular, as dangerous. Certainly there are dangerous reptiles, snakes among them, but they are the exceptions, and one's chances of encountering such an animal in a rainforest are remote. Still, they do exist, and they are worth considering.

Venomous snakes occur in all rainforests, as they do throughout the tropics. Indeed, some of the most venomous are found in desserts, grasslands, and deciduous forests. Venom evolved in snakes as a means of subduing prey and secondarily as defense. However, even most venomous snakes would rather run than fight, and people who get bitten by rainforest snakes usually have either cornered or provoked the snake, such as accidentally stepping on one.

In my experience in Central America, the poisonous snake most frequently encountered in the rainforest is the eyelash viper, also known as the palm viper. This is an arboreal animal, rarely coming to the ground and capable of clinging to the vertical trunks of trees. The typical coloration of the snake blends in with the vegetation quite well. A person carelessly grabbing a tree while walking could be in for a painful surprise should he put his hand on such a viper. Supposedly, the eyelash viper is ill tempered and will strike even if not provoked. My experience does not support this; however, I've not tried getting close enough to one to test the idea. Interestingly enough, there are a number of color variants, including bright yellow-gold and pink. The bite of an eyelash viper can conceivably kill a human, particularly since most bites are on the upper half of the body.

Rainforest destruction and expanding human populations may be responsible for more human and snake interactions and, consequently, more cases of snakebite. In Costa Rica, for example, abandoned macadamia nut plantations originally cut out of the forest have turned out to be hatcheries for the fer-de-lance, a deadly tropical pit viper that has been drawn to the plantations by rats, which were drawn, in turn, but the abundance of nuts. Both snakes wandering from the plantations into residential areas or people going to the plantations to gather nuts bring snakes and people together, and the result can be injurious.

Poisonous snakes are to be avoided, by human and animal alike, and anything that encounters one and survives is likely to try very hard to avoid repeating the experience. One born with instinctive avoidance behavior would be at least as likely to stay away. Consequently, any *nonpoisonous* snake that resembles a poisonous one should be at an advantage. A case in point is the poisonous coral snake (Figure 7-5) and the nonpoisonous false coral snake. Actually, there are several species of each, but for the purpose of this discussion, we'll use the "typical" American coral snake, which ranges from the southern United States south through Mexico, Central America, and well into South America. It is not restricted to the rainforests, but it does occur in them. This animal is characteristically small, usually less than two feet in length and one inch in width, and is brightly colored with red, yellow, and black rings,

Figure 7-5 True coral snake.

with the red and yellow rings adjacent to one another. The false coral snakes, which coexist over much or the same range, are colored differently. For example, in Costa Rica false coral snakes are red and black ringed, with no yellow. Elsewhere, false coral snakes may have yellow rings that do not touch the red ones. The folk wisdom for distinguishing between true and false coral snakes is some variant of the phrase, "Red and yellow; kill a fellow. Red and black; you're okay, Jack." Unfortunately, there are some true tropical coral snakes that don't conform to the general rule, but for the most part, the false coral snakes still benefit by their resemblance, however perfectly or imperfectly, to the poisonous models. Research conducted by David Pfennig and William Harcombe at the University of North Carolina has shown that where their ranges coincide, false coral snakes are attacked much less frequently by potential predators than where the mimic exists but the model is lacking.[11]

The phenomenon described in the previous paragraph is not uncommon in nature. A similar and well-known case of such mimicry occurs in North America between unpalatable monarch butterflies and the palatable viceroy butterflies that mimic them. This kind of mimicry was first described by British naturalist Henry Bates in the mid 1800s. It is now known as **Batesian mimicry**.

The coral snakes discussed above have numerous relatives in the Old World, including the mambas of Africa, some of which are rainforest dwellers; the cobras of Africa and Asia, again some of which live in rainforests; and a variety of snakes in Australia. Other poisonous snakes exist, of course. In tropical America, these include principally pit vipers such as the tropical rattlesnake, the bushmaster, and the fer-de-lance and palm vipers discussed earlier. Pit vipers also occur in Asia but neither in Africa nor Australia. Tropical Africa has some pretty nasty snakes in addition to the mambas and cobras. The Gaboon viper, for example, a true viper that lives in the African rainforest, has the longest fangs and delivers the greatest volume of poison of all venomous snakes.

Another group of snakes that seem to fascinate people are the large constrictors: the boas and anacondas of the Americas and the pythons of the Old World. These animals subdue their prey by looping coils of their bodies around the prey and crushing them. There are conflicting stories over whether anacondas or pythons, both of which reportedly reach lengths over 30 feet, are the largest of the snakes. The reticulated python of Asia is the longest, but the golden brown anaconda of Amazonia is the most massive. Stories of some of these eating people exist, but are rare. Oddly, some people elect to keep pythons as pets. However, when these pets have grown to be too large to comfortably keep, some of those owners have released them. In Florida, and possibly other places in the United States, pythons can and do survive once released, and they can cause problems to native wildlife.[12]

When invasive animals such as pythons are introduced into a new, suitable habitat, they often find favorable conditions due, in part at least, to an absence of predators. Moreover, potential prey have not "learned" that the exotic is dangerous. This is illustrated by the brown tree snake (Figure 7-6), a native of northern Australia, Papua New Guinea, and other islands in the region. It lives in the rainforest and other tropical ecosystems. The snake, generally nocturnal in habit, has spread to many Pacific islands by hitching a ride in airplane cargo and wheel wells, where it may go to avoid daylight. It is well adapted to island rainforests, and it has prospered in its new homes, having devastated bird and small mammal populations in the process, the island of Guam being a principal example. In the cases of birds and bats involved in seed spreading and plant pollination, their declining populations due to the snake are likely to result in a cascade of declining populations of plant species as well.[13]

Other tropical reptiles include lizards, turtles, and crocodilians. Among these, the iguana is usually and easily observed; it is large, reaching 6 feet in length, and common. It is a vegetarian, feeding principally on fruit and leaves; consequently it is abundant as well, which makes it a good potential protein source, particularly for natives. The Bribri Indians in Costa Rica raise Iguanas for this purpose. Small lizards are also very abundant. Basilisks, in particular, can often be seen scurrying across the forest floor and even running on the surface of streams, thus giving them the name "Jesus Christ lizard."[6]

Figure 7-6 Brown tree snake.

The monitor lizards in tropical Australia, Africa, and Asia are large, predatory reptiles. On many of the Indo-Pacific islands where no large, predatory mammals occur, these may be the top carnivore. A case in point is the Komodo dragon, found on some Indonesian islands, which can weigh over 300 pounds. In Africa, there are two kinds of monitors, one of which, the Nile monitor, is found in the rainforest. It reaches a length of 6 feet.[14] Not all monitor lizards are rainforest animals, and not all of them are large. Undoubtedly, these animals have a long evolutionary history and are quite adaptable.

In the Americas, alligators and caimans make up the most common crocodilians, with caimans actually being the more tropical. They are found throughout Central and South America, although they are being threatened by overhunting and habitat destruction. In Africa and Asia, true crocodiles are the top predators of the waterways. American crocodiles exist, but they are rare. The largest African crocodile, the Nile crocodile, reaches a length of 16 feet. It is distributed through much of tropical and southern Africa. The largest crocodile of all is the Indo-Pacific crocodile, which lives in salt and brackish water throughout tropical Southeast Asia and grows as long as 23 feet.[15] Perhaps mainland populations of this crocodile may encounter tigers or leopards or even a large python on occasion, but other than humans who hunt it for its hide, it is the biggest carnivore over much of its range, particularly in Australia and the Pacific islands.

Amphibians

These are sometimes spoken of as transition vertebrates, having bridged the water-dwelling fish and terrestrial higher vertebrates. Amphibians characteristically lay their eggs and spend the juvenile component of their lives in water and become terrestrial as adults. In the American tropics, there is a large diversity of frogs and toads, but salamanders, in contrast, are rare. Among the frogs, however, are some of the most colorful animals in the rainforest. These are the poison dart frogs (Figure 7-7), and their bright colors are essentially warnings to other animals that they are toxic. Ironically, these frogs lose their toxicity in captivity; they become toxic in the wild because

Figure 7-7 Poison dart frog.

of their diets, mostly small insects. Obviously, they are immune to toxins their prey contains. The toxins are located on the animals' skin; a predator does not have to eat the frog to be affected by it. Holding the frog in its mouth is generally enough. Generally, the toxins do not penetrate human skin; you can usually handle these frogs safely enough if you have no cuts on your hands. If you do, or if you inadvertently lick the hand that held the frog, however, you may be in serious trouble. Natives of the South American rainforests are known to use the frogs' skin secretions on their blowgun darts, hence the animals' name. They also show some unusual reproductive behaviors for amphibians, such as laying eggs on leaves over water into which the tadpoles will fall, or transporting the tadpoles to bromeliads in the canopy once the eggs develop.[6]

Fish

To many people, tropical fish are small, brightly colored, and swim around aquaria. As with most animals, however, fish are quite diverse in the tropics. The tetras, for example, that are popular among aquarists are characins, a fish family of South America that includes piranha. The tetras and catfishes make up around two thirds of Amazon species.[6] Fish diversity would seem to have been carried to an extreme by the fish family Cichlidae, the cichlids. These are fish that are found in tropical America and have made their way into the tropical fish trade. A few species are found in India as well, but the greatest diversity appears to have occurred in Africa. Lake Victoria, not within the current limits of the African rainforest, contained 300 or more species, and the nearby Rift Valley Lakes, according to E. O. Wilson, swarm with them.[16]

The Amazon and its tributaries contain some of the most fascinating fish, if from a somewhat jaundiced point of view. Fish like the piranhas, which reputedly can strip a human body to bone in minutes, and electric eels, which are capable of delivering a shock of greater than 600 volts, would be things to avoid. More bizarre is a group of catfishes known as candiru that includes a species that supposedly will swim into human orifices. A very thin fish, it is a gill parasite of other fishes, but it is supposedly attracted by urine. It is said to swim into the urethra of a human victim, flare its spines, and become lodged, living off its victim's blood. According to Wheaton College biologist John Kricher, that and the notion that piranhas will spontaneously tear into mammalian prey, like the man-eating anaconda, may be apocryphal.[6] Catfish are a diverse group of fish, and some get to be huge, such as the Mekong giant catfish that is found in Southeast Asia.

One more oddity among Amazon fishes is that some of them are specialized for consuming terrestrial fruit and seeds. Those that live among flooded trunks in igapo forests help in the dispersal of seeds that pass through the fish's' intestines when fruit is eaten. Even piranha will eat seeds when they are abundant.[6]

As with snakes, tropical fish have been imported into North America as pets and have been released or escaped, sometimes with adverse consequences. Others have been transplanted with good intentions but have ended up as problems. Piranhas have shown up in Florida and elsewhere, and the Nile perch, deliberately planted in Lake Victoria as a food and game species has wreaked havoc on native cichlids, possibly causing extinction of some species.[16] One of the more bizarre exotics that has established itself in the United States is the so-called walking catfish, a fish that is able to literally climb out of a pond and make its way over land to another. In Florida it has spread into irrigation ditches and other waterways.[17]

In Closing

The birds, reptiles, amphibians, and mammals show much diversity in rainforests because they've evolved to fill a variety of environmental opportunities. Some types have evolved into strange, even bizarre species, and some have been able to attain large sizes. Undoubtedly the year-round growing season and variety of foods available to them has been a factor in this diversity, which, in all probability, has not yet been fully realized.

References

1. Peterson, R. T. 1990. *Peterson field guides: Western birds*. Boston: Houghton Mifflin.
2. Carlson, R. 2005. Bountiful Panama or how to see 460 species in two weeks. Tennessee Ornithological Society Report from Panama. http://www.advantage panama.com/ Panama-Birds-Tennesse-Audubon-Report.htm.
3. Pickrell, J. 2005. Terror birds: Predators with a kung fu kick? *National Geographic News*. http://news.nationalgeographic.com/news/2005/08/0801_050801_terror birds.html.
4. Roberson, D. 2000. Parrots: Psittacidae. http://montereybay.com/creagrus/ parrots.html.
5. Roberson, D. 2003. Kingfishers: Alcedinidae. http://montereybay.com/creagrus/ kingfishers.html.
6. Kricher, J. 1999. *A neotropical companion*. Princeton, NJ: Princeton University Press.
7. Ramel, G. 2001. Hornbills: The Bucerotiformes. http://www.earthlife.net/ birds/ bucerotiformes.html.
8. Bildstein, K. L. et al. 1998. Conservation status of tropical raptors. *Journal of Raptor Research* 32(1):3–18. http://hawkmountain.org/media/054%20Bildstein%201998. pdf.

9. The Peregrine Fund. http://www.peregrinefund.org/.

10. Bueser, G. L. L. et al. 2003. Distribution and nesting density of the Philippine eagle *Pithecophaga jefferyi* on Mindanao Island, Philippines: What do we know after 100 years? *Ibis* 145(1):130–135.

11. Smith, J. 2001. A striking resemblance: Con artists that crawl. *Endeavors, the University of North Carolina at Chapel Hill.* http://research.unc.edu/endeavors/fall2001/pfennig.htm.

12. Lovgren, S. 2004. Huge, freed pythons invade Florida Everglades. *National Geographic News.* http://news.nationalgeographic.com/news/2004/06/0603_040603_invasivespecies.html.

13. Patrick, L. 2001. Introduced species summary project: Brown tree snake (*Boiga irregularis*). Columbia University Information Technology. http://www.columbia.edu/itc/cerc/danoff-burg/invasion_bio/inv_spp_summ/boiga_irregularis.html.

14. Honolulu Zoo. Undated. *Monitor lizards.* http://www.honoluluzoo.org/monitor_lizards.htm.

15. Nova. Undated. Who's who of crocodilians. PBS.org. http://www.pbs.org/wgbh/nova/croccaves/who-nf.html.

16. Wilson, E. O. 1992. *The diversity of life.* Princeton, NJ: Princeton University Press.

17. Robins, R. H. Undated. Walking catfish. Ichthyology at the Florida Museum of Natural History. http://www.flmnh.ufl.edu/fish/Gallery/Descript/WalkingCatfish/WalkingCatfish.html.

Arthropods

8

ALTHOUGH THE VERTEBRATES MAY BE the charismatic animals in the forest, the invertebrates are by far the most abundant, in terms of numbers, diversity, and biomass. In particular, the arthropods are the biggest single phylum of animals, and they are the ones visitors to rainforest will most frequently encounter. Terrestrial mollusks, worms, and other invertebrates are present but are not nearly as noticeable nor, probably, as abundant or diverse. In fact, there may be more different kinds of beetles, one class of insects, than all kinds of vertebrates, and the biomass of just the ants on the planet may be equivalent to that of humans.[1] Anyone who wants to discover a new species would be well advised to look among the invertebrates, particularly the arthropods and especially the insects.

Arthropods crawl and fly above and through the canopy, up and down the trees, through the understory, across the ground, and in and out of human dwellings. A substantial number burrow into the ground. They buzz, hum, and click; they bite, sting, and scratch. A few are poisonous, and some are disease vectors. But some are critical in the pollination of flowers, and others are vital in cleaning up dead and decaying matter. A few, such as the brightly colored butterflies, are even attractive. Additionally, many serve as food, directly or otherwise, for the charismatic animals, a few even for us, and many are responsible for the destruction of some of the more obnoxious members of their own biotic categories. In short, regardless of what one may think of them, the invertebrates are a critical part of the tropical environment. Without them, the rainforest could probably not exist, at least not in its current form, and the world, as we know it, would be radically different.

The Insects

Because of their numbers, importance, and diversity, the insects are a logical assemblage to start this discussion with. There are approximately 950,000 known species of insects and perhaps more than 7 million yet to be described—equating to more than half of all known living things.[2] Even dealing with only those that inhabit the rainforests would be an impossible task in a book such as this. Therefore, the descriptions that follow represent only a sampling of the variety that exists.

The Ants

Collectively, ants probably have a greater impact on their respective ecosystems than any other animal, humans often included. Living in colonies of thousands or hundreds of thousands, they move more earth, channel more energy, and impact more other species than any other type of animal, and they occupy an amazing variety of niches. Along with termites, they account for a third of Amazon biomass.[3]

Among the most fascinating members of this category are the army ants, a group that has no permanent home but lives in moveable concentrations of intertangled workers' bodies called bivouacs. Every day, workers surge out of the bivouacs on raids. When they come across another animal, they attack in mass, quickly killing it and cutting it into pieces to be returned to the bivouac and consumed. Mythology has it that these will eat anything that gets in their way, humans included, but in reality, they rarely attack anything larger than a mouse. However, they do have strong mandibles, and if they climb on you, you are very likely to get bitten as you brush them off. Even so, scorpions, tarantulas, small lizards and snakes, and virtually every other small animal is vulnerable and well advised to get out of the way when a swarm is coming. That, however, is no guarantee of safety as certain species of birds, known appropriately enough as antbirds, follow the swarms and feast on invertebrates that the ants flush out. [4] Sometimes a swarm will invade a house, in which case a room or two will be totally cleaned of unwanted residents within a couple of days. The owners of the house simply avoid whatever rooms the swarm is occupying until it leaves on its own accord. The bivouac period corresponds to the queen ants' period of laying eggs. Once she has finished, the colony moves to another bivouac and the cycle begins again. Army ants are not restricted to tropical forests, but they are abundant in them, and they are found in Africa, where they're known as driver ants, and in Asia as well as in tropical America. [1]

Not all ants are predatory. A number eat vegetable material, such as seeds, and some live on fungus, which they grow themselves. The best example of the latter is the leaf cutter ants of the neotropics. These ants grow their subterranean fungal gardens with pieces of leaves they've clipped off plants. Scouts find trees to the species' liking and return to the colony laying down a chemical trail in the process. Workers follow the chemical trail back to the tree, cut sections of the leaves, and carry them back to the colony. In time, the constant stream of workers to and from the tree wears a miniature trail in the forest floor, and one can observe the ants marching robotically between the two points (Figure 8-1). Back at the colony, the leaf fragments are passed on to smaller workers, who chew the leaf material into a pulp, sometimes adding an anal secretion, and pack the pulp into a wad. A bit of fungal culture is added to the wad, and the culture takes off from there. The ants feed the fungus to their larvae, and they eat some themselves, but they may also supplement it with plant sap. [3,5] Leaf cutter colonies can become enormous, containing millions of individual ants and occupying a square meter or more of ground space. The fungus that the leaf cutters grow is not found outside of the colony. Consequently, it is as dependent upon the ant for its survival as the ant is on it. Such mutual dependence suggests a long association between the two organisms where the two have evolved together into their present forms.

Bees and Wasps

Within the same order as the ants, bees, and wasps are also extremely important members of rainforest fauna. Their principal value is their role as pollinators, but a number of wasps are parasitic and may be important as control agents for pests. It was once thought that parasitic wasps were actually less common in the tropics than in temperate zones, but one recent study has

Figure 8-1 Leafcutter ants.

reported that at least in the Americas, this is not the case.[6] Moreover, the relationship between pollinating wasps and the plants they pollinate represents more indicators of coevolution and mutual interdependence. Some of these are discussed in the next chapter.

Bees are general pollinators; they tend to visit a variety of plants. Among the more characteristic bees of the American tropics are the euglossine bees, brightly colored metallic blue and green individuals, in which it is the male that does the pollinating. Unlike honeybees, euglossines are more solitary, and the males appear to try to attract females with attractive odors or with chemicals they manufacture from precursors, some of which they may collect from flowers.

In general, bees and wasps mind their own business, and someone who gets stung has usually somehow disturbed the insect, admittedly usually unintentionally. Many bees and wasps probably restrict their activity to the canopy where most of their food is found. I've seen impressive hornets' nests hanging from high branches in Central America, but, oddly enough, I've yet to see a hornet there. I suspect they don't come down into the understory. However, there is one type of bee that has become a problem. About 50 years ago, Brazilian scientists attempted to improve honey production in their country by crossing domestic honeybees, originally imported from Europe, with queen African honeybees. The goal was to breed a bee with the pollinating and honey-making capabilities, along with the comparatively tractable disposition, of the European bee but having the heat tolerance of the African bee. The result could not have been a better illustration of Murphy's Law. The bees that were bred turned out to have virtually none of the desirable characteristics of the European bee but all of the disagreeable traits of the African. To make matters worse, a number of African queens escaped and began hybridizing with European bees, the result being the so-called Africanized or killer bees. Since then, the Africanized bee has continued to hybridize with feral honeybee colonies as well as reproduce on its own and has spread throughout South and Central America as far north as Texas. Individual hives have been found even further north and east. Moreover, there has been surprisingly little or no attenuation of the Africanized bees' unattractive qualities. Theoretically, their advance north should be slowed as they start to encounter North American cold winters. An individual Africanized bee is difficult to distinguish from a normal honeybee. They look virtually totally alike, and their stings are essentially identical. However, individual Africanized bees are far more nervous, and a colony could be described as psychotic. Africanized bees are far more likely to mount an unprovoked attack if someone ventures too close to their hive; they will chase a supposed intruder as much as a quarter of a mile. When they do attack, their attacks are much more vicious with many more bees getting involved. This was learned the hard way when initial attempts to control the bees were by someone taking an axe to a hive. In the resulting melee, the attacker was stung to death, as have been perhaps 1000 other Brazilians over the past 50 years. Needless to say, Brazil's honey production has declined. In several trips to the Central American rainforest, I have never seen Africanized bees in the rainforest per se, although it's hard to imagine that they were not present. But, I have seen them on agricultural land that was rainforest before it was cleared, and given the rate of rainforest destruction and the growing problem of global warming, it's probable that they will become more of a problem.[7] In contrast, the Africanized bees have moved into regions of the Amazon that European honeybees were unable to occupy, and apparently they have become something of a resource, both in terms of providing pollination in reforested areas where native bees have become extinct and as producers of honey.[8]

Termites

This is another cosmopolitan insect, not restricted to the tropical rainforest but certainly common within it. Termites have their greatest distribution in Africa, but they are present in rainforests around the world, where they are important in the decomposition of dead vegetation. Their mounds can be seen abundantly in Central America. Tropical termites do not infest houses, as do temperate ones. They live in large mounds, actually some reaching two meters in height in Africa and Australia (Figure 8-2). In Africa they also serve as a protein source for chimpanzees and other primates. Termites have been implicated by some as prime contributors to global warming because of their release of the greenhouse gasses carbon dioxide and methane, but their role, in fact, in global warming is overstated. The carbon dioxide they produce is simply part of the global carbon cycle. If it were not produced by termites, some other decomposer that breaks down plant material would produce it. Their contribution to global methane production is also a matter of debate, as estimates vary widely.[9] I would guess that it's substantially less than the contributions of domestic cattle, and according to Songwe et al., they are very important in the overall processing of leaf litter and wood fragments and the subsequent release of nutrients into forest soils.[10]

Butterflies

Every once in awhile, everyone has an experience that he or she remembers for his or her entire life. I've had several in the rainforest, but one that stands out in particular occurred at the base of the La Fortuna waterfall in Costa Rica. The waterfall is not particularly hard to get to; in fact, it's a popular tourist spot, and one would be hard pressed to not find native Costa Ricans and tourists picnicking near the base and swimming in the clear pools downstream. The waterfall is impressive, plunging more than 200 feet (70 m) or more into a deep, narrow canyon. One morning, I left the group I was leading in the capable hands

Figure 8-2 Termite mound.

of a Costa Rican guide and descended into the canyon alone, arriving at the waterfall base before anyone else that day. A mist was rising from the base of the falls, nothing unusual, and enough sunlight was filtering down through the canyon, very unusual, to have caused a rainbow to form over the pool. Under the arc of the rainbow, two blue morpho butterflies flitted in and out of the mist like two brilliant, iridescent blue sprites. The scene was breathtaking, even spiritual.

The blue morpho is one of the largest butterflies in Costa Rica and my personal favorite, despite that it gets its nourishment by sucking juices from decaying fruit or animal carcasses. However, it is only one of many species that you can find in the tropics. The closer you get to the equator, the more diverse the butterflies become, reaching their zenith in western Brazil and in Peru.[11] Most tropical butterflies have a more glamorous lifestyle than blue morphos; they sip nectar from and pollinate flowers. Ironically, many of their caterpillars are destructive to plants, which brings up an interesting note. Rainforest plants often produce toxic chemicals, a trait that is supposed to discourage herbivorous caterpillars from eating their leaves. However, some species of butterflies, or more appropriately their caterpillars, have "learned," in the evolutionary sense, to handle the toxins. Some are able to metabolize them, while others, such as members of the genus *Heliconius*, sequester the toxin somewhere in their bodies where it does them no harm, as happens in the case of the North American monarch butterfly caterpillars. Any animal that tries to eat one is usually sickened by the toxin and learns to leave the caterpillars alone. The toxin persists in the tissues of the adult butterfly, and once more, any animal, usually a bird, which tries to eat one, gets sick and from then on leaves the species alone. Earlier, the idea of Batesian mimicry was discussed using the false and true coral snakes as examples. Among the heliconid butterflies of the Americas, there are several toxic pairs of species where one member resembles the other. A bird eating the model or the mimic, which would be academic in this case, will become ill and avoid both species thereafter. Mimicry where two foul-tasting species resemble each other is known as **Mullerian mimicry,** named for Fritz Muller, the German biologist who first described it.[4] Mullerian mimicry would seem more efficient than Batesian in that a predator who had first encountered the mimics would not have to "unlearn" how to eat them.

Notwithstanding the ecological importance and adaptations of the many tropical butterflies, whether it's the Ulysses of Australia, the Raja Brooke of Malaysia, the African mocker swallowtail, or any of the other thousands of species found throughout the rainforests, these are beautiful organisms. Anyone's trip to the rainforests would only be enhanced by his or her seeing them.

True Flies

These are the last insects to be considered in this chapter, and the importance of some of them lies not in their beauty or ecological roles, but in their impact on human and animal health. Many of them are vectors of infectious disease, and some are parasitic as larvae, thereby being disease agents in their own right. To be fair, some flies are important as carrion feeders or as pollinators, and some parasitize or prey upon other insects. Others still serve as food for more desirable insects, as well as birds and amphibians. But in general, from the human perspective these are not the most alluring members of the insect class.

Flies are distinguished from other insects by having only a single pair of wings. Most other insects, regardless of whether or not their name includes the word fly, have two pairs of wings. Among the most obnoxious and dangerous of these animals are the biting flies, because they are capable of and frequently do transmit disease organisms from one of their victims to another.

Perhaps the best example of these, and a good starting point, would be the mosquitoes. The list of diseases carried by this particular fly is long, and it is restricted to neither the tropics nor to rainforests. However, such diseases as dengue fever,[12] malaria, and yellow fever[13] are considered to be tropical in origin. Lymphatic filariasis, sometimes called elephantiasis, is considered to be entirely tropical.[14] Malaria has become seriously problematic throughout much of the developing world. As of 2002, malaria was the fourth leading cause of death worldwide, killing more than a million people a year. Most of the deaths occurred throughout much of sub-Saharan Africa, not only in the rainforest, which is not the hardest hit. However, it is present in the forested parts of Southeast Asia, Amazonia, and Central America.[14] One would expect that the rainforest would be alive with mosquitoes and that malaria would be a threat to anyone visiting it. It is, however, more complicated than that, and my experiences have been otherwise. First, malaria can be spread by only some species of the mosquito genus *Anopheles*, and those species must be present. Secondly, a reservoir for the malaria parasite, usually infected people, must be present at well. I have encountered mosquitoes in the rainforest; I would imagine everyone who has gone there has as well. But as with other animals, there is a much greater diversity of mosquitoes in the rainforest; consequently, there are genera other than *Anopheles* that compete with it and may possibly keep its numbers down. Furthermore, there are greater numbers of mosquito predators. This is not to say that the threat of malaria, or any other mosquito-borne disease, is overstated, it is simply not a problem universally through the rainforests. Moreover, there are steps one can take to protect oneself, such as using mosquito repellant, taking quinine prophylactically, and staying indoors while mosquitoes are most active. Avoiding swampy areas where mosquitoes are more abundant is also a good idea.

Three other biting flies and the diseases they spread are black flies and river blindness, tsetse flies and African sleeping sickness, and sand flies and Leishmaniasis. Again, these are not restricted to rainforests; if and when the rainforests completely disappear, these infections will continue to plague parts of the world, with sleeping sickness and river blindness predominantly if not entirely in Africa and Leishmaniasis predominantly in Southeast Asia but also in the Americas.[15]

Once more to be fair, organisms other than the flies serve as disease vectors. One in particular, Chagas disease, a protozoan endocarditis spread by a type of true bug, is largely found in the American tropics and may have infected Charles Darwin. However, flies, or more appropriately the illnesses they spread, have been around for a long time. They have caused much human suffering and have even impacted human history. Attempts to eradicate them have largely failed and may in some cases have made things worse. As noted above, rainforest destruction has not alleviated matters, as many of the organisms that spread the diseases either are capable of living outside of the rainforests or have close enough relatives that do. Both poverty and a growing human population continue to keep people in conditions that promote the spread of these diseases. Until and unless that changes, tropical diseases will be a problem.

One last fly to mention in not a vector for parasites; it *is* the parasite. This is the botfly, a collection of species that parasitize mammals, including ourselves. The botfly often attacks indirectly. An adult female will seize a mosquito or possibly other kind of biting fly, lay some eggs on it, and then release it. When the mosquito visits a mammalian victim, body heat hatches the botfly egg, and the larva falls onto and burrows into the mammal's skin and feeds on its host's tissues until it matures and leaves. A single infection on a human is generally a painful irritation but nothing serious, and it's easily treated. However, multiple simultaneous infections can be debilitating. The botfly is found in tropical America. Infection is often unreported among natives; it's usually a tourist that pays attention should he or she get infected. However, in Central America, the botfly is also a serious pest of cattle, causing financial difficulty for ranchers. Similar pests exist in Africa.[16,17] Moreover, it can also infect native animals, where it can cause acute problems. A single fly larva in a 25-pound monkey has a much greater impact than it does in a 150-pound human, making the monkey sick and susceptible to other infections. The expansion of cattle ranching in the tropics has broadened the food base for botflies, thus making them more of a potential problem. Much of the beef produced there ends up in fast-food hamburgers in North America. You might think about the botfly next time you're deciding whether or not you want fries with that.

Arachnids: Spiders and Scorpions

On one of my earlier student trips to Costa Rica, we spent a night near the Arenal volcano in the northern lowlands of the country. We had just gotten settled into our cabins when the air was split by a blood-curdling scream. Five of us, our Costa Rican guide included, stormed the cabin and found one of the girls occupying it standing on her bed, one arm across her body, pointing at a big, hairy spider on the floor, and crying, "Kill it! Kill it! Kill it! Kill it!" The spider was never identified; the first guy into the cabin flattened it with his shoe. On another trip, I spent about 15 minutes of the first night at a group session before turning in imploring the students to thoroughly shake out their clothes and shoes before dressing in the morning. The following morning, our guide came to breakfast with a scorpion in a drinking glass. "*I should have listened to you,*" he said. "This (Spanish expletive deleted) hit me three times." He discovered the scorpion in his jeans that morning when he pulled them on.

Scorpions and big spiders are scary; they make me squeamish too, but by and large, they're not dangerous, although after seeing the movie *Arachnophobia*, I wasn't completely convinced. In general, they prey on other small animals, mostly insects, but there is a tarantula in northern South America that's big enough to eat small birds (Figure 8-3), which it sometimes does. With a leg span of about a foot (30 cm), the Goliath bird-eating spider can subdue small birds when it can catch them. Like all spiders, it is venomous, but like all tarantulas, its venom is almost always not toxic to humans, and it bites only in defense. Their body hair can also be thrown at a possible at-tacker, and that is irritating. It's probably not a good idea to try to pet one of those things.[18] Tropical spiders exist in abundance, and some of them are very impressive in size, but there are far more dangerous things out there.

Scorpions (Figure 8-4), while not spiders, are closely related to them. They are a very old and primitive form of life, and they are abundant through-out tropical and semitropical ecosystems. They typically prey upon other arthropods, which they seize in their crablike pincers and subdue with their sting. They also use their sting when they feel threatened, as my Costa Rican friend had demonstrated. Scorpions are nocturnal hunters, preferring to re-main reclusive during the day. At night they stalk their prey, and for some rea-son, they like to crawl into clothing and shoes, perhaps thinking that these are good places to hide. Unlike tropical spiders, however, some scorpions are dangerous, even deadly. The most toxic are to be found in North Africa

Figure 8-3 Goliath bird-eating tarantula.

and the American Southwest—both desert areas.[19] According to Gilbert Calvo, a Costa Rican guide and naturalist, the sting of rainforest scorpions is not harmful, while those from the drier areas of his country are. However, it is still probably a good idea to remain safely in bed at night, wherever in the tropics

Figure 8-4 Scorpion.

one may happen to be, and to shake out one's shoes and clothing in the morning.

Crustaceans

Although insects are the most successful animals on land, crustaceans, the crabs and similar organisms, are the most successful in general in water. However, just as there are aquatic and even salt-water insects, there are terrestrial crustaceans. Most people are acquainted with hermit crabs, which you can often see scuttling across beaches in the tropics, but other tropical land crabs exist, including some that live in trees. Mangrove forests all over the world have crabs living on their exposed roots and in their canopies, where they may eat leaves. Tropical islands often have crabs actually living in the forests. Christmas Island, in the South Pacific, for example, has a species of land crab that lives on the forest floor and eats fruits and seeds. It is known to sequester some seeds in its burrows, which may be factors in new plants growing when gaps in the forest occur.[20] However, tree-climbing crabs exist in continental forests as well. One species in particular occupies water-filled tree holes in the Tanzanian rainforest. In the canopy it hunts snails, which it eats, but it also uses the snails' shells to provide calcium for its exoskeleton. It dissolves the shells in the calcium poor water of its tree holes.[21]

In Closing

The animal diversity within the tropical rainforests is immense. It would be impossible to cover it in a book of this nature. Consequently, the animals that have been discussed in the last three chapters are only representatives of a vast zoology, much of which is yet to be uncovered and described. Every tropical island in the Indian and Pacific Oceans most likely has its own varieties of tree crabs, and the known insects of tropical America can fill volumes. What value many of these creatures may have as sources of food and medicines are subjects for little more than speculation right now. Finding out will require the preservation of the forests in which they live. That will be a subject for later chapters.

References

1. Hölldobler, B., and E. O. Wilson. 1994. *A Journey to the ants: a story of scientific exploration.* Cambridge, MA: Belknap Press.
2. United Nations Environment Programme. 1999. GEO-2000: Biodiversity. *Global Environmental Outlook.* http://www.unep.org/geo2000/english/0045.htm.

3. Hölldobler, B., and E. O. Wilson. 1990. *The ants*. Cambridge, MA: Belknap Press.
4. Kricher, J. 1997. *A neotropical companion*. Princeton, NJ: Princeton University Press.
5. San Juan, A. 2005. *The lurker's guide to leafcutter ants.* http://www.blueboard.com/leafcutters.
6. Sääksjärvi, I. et al. 2004. High local species richness of parasitic wasps (Hymenoptera: Ichneumonidae; Pimplinae and Rhyssinae) from the lowland rain forests of Peruvian Amazonia. *Ecological Entomology* 29:735-743. http://vanha.sci.utu.fi/amazon/The_team/Publications/individual_publications/Saaksjarvi_et_al_2004_EcolEnt.pdf.
7. Department of Systematic Biology, Entomology Section, National Museum of Natural History. Undated. Buginfo: Killer bees. *Smithsonian Institution Information Sheet #45.* http://www.si.edu/Encyclopedia_SI/nmnh/buginfo/killbee.htm.
8. Roubik, D. W. 1992. *Ecology and natural history of tropical bees*. Cambridge, UK: Cambridge University Press.
9. United States Environmental Protection Agency. 1996. Termites—Greenhouse gasses. http://www.epa.gov/ttn/chief/ap42/ch14/final/c14s02.pdf.
10. Songwe, N. C., D. U. U. Okali, and F. E. Fasehun. 1995. Litter decomposition and nutrient release in a tropical rain forest, Southern Bakundu Forest Reserve, Cameroon. *Journal of Tropical Ecology* 11:333–350.
11. Wilson, E. O. 1992. *The diversity of life*. Cambridge, MA: Belknap Press.
12. National Center for Infectious Diseases. 2004. *Infectious disease information: Mosquito-borne diseases*. Centers for Disease Control and Prevention. http://www.cdc.gov/ ncidod/diseases/list_mosquitoborne.htm.
13. Diamond, J. 1997. *Guns, germs, and steel: the fates of human societies*. New York: W.W. Norton & Co.
14. Marcus, B. A. 2004. *Deadly diseases and epidemics: Malaria*. Philadelphia: Chelsea House.
15. Centers for Disease Control, Division of Parasitic Diseases. http://www.cdc.gov/ncidod/dpd/index.htm.
16. Hunter, J. M. 1990. Bot-fly maggot infestation in Latin America. *Geographical Review* 80(4):382–398.
17. Westmead Hospital. Undated. *Exotic myiasis*. Westmead Australia, Department of Medical Entomology, Westmead Hospital. http://medent.usyd.edu.au/fact/myiasis.html.
18. Benders-Hyde, E. M., Ed. 2006. Goliath bird-eating spider. *Blue planet biomes*. http://www.blueplanetbiomes.org/goliath_bird_eating_spider.htm.
19. Speer, B. R. 1996. *Scorpions*. University of California Museum of Paleontology. http://www.ucmp.berkeley.edu/arthropoda/arachnida/scorpiones.html.
20. O'Dowd, D. J. and P. S. Lake. 1991. Red crabs in rain forest, Christmas Island: Removal and fate of fruits and seeds. *Journal of Tropical Ecology* 7(1):113–122.
21. Bayliss, J. 2002. The east Usambara tree-hole crab (Brachyuara: Potomoidae: Potamonautidae) – A striking example of crustacean adaptation in a closed canopy forest, Tanzania. *African Journal of Ecology* 40(1):26–34.

Biotic Interrelationships

IN 1862, CHARLES DARWIN, WHILE studying the Madagascaran orchid *Angraecum sesquipedale* that has a foot-long nectary, predicted the discovery of a moth with an 11-inch proboscis that functions as the orchid's pollinator.[1] Fifty years later, the moth *Xanthopan morganii praedicta* was indeed discovered. It actually had a 12-inch proboscis, but it did sip nectar from and pollinate *Angraecum sesquipedale*.[2] It would look like the relationship benefits both partners. The moth has a source of nectar for which it has no competition, and the orchid has a dedicated pollinator.

Wherever the ambient temperature rises above freezing for any period of time, there appear to be plants. In the arctic and alpine tundras, where the growing season is less than two months, stunted plants cover the landscape. Most reproduction is vegetative, but some sexual reproduction occurs, and a few of the plants rely on an insect pollinator. Almost without exception, that pollinator is a bumblebee that visits whatever flowers happen to be blossoming. Since plant diversity is low, there is a good chance that the bee will carry pollen from a given flower to a conspecific. In the tropical rainforests, however, with so many kinds of plants blossoming simultaneously and often a considerable distance existing between conspecifics, it's conceivable that a bee might visit a given species only once on a foraging trip. At the very least, it's likely that the bee will not carry one plant's pollen directly to a receptive flower, and whatever pollen the bee happens to pick up from a particular species ends up being wasted. Consequently, it is to a plant's advantage to have a dedicated pollinator, some kind of pollen vector that anatomically, physiologically, or behaviorally is obligated to visit flowers of a given species one after another. In some cases, **coevolution**, two separate species evolving together, may cause two species to become biologically inseparable. A plant and its pollinator would evolve into mutual dependence. Should anything happen to one, however, the other is doomed. In the case of Darwin's orchid and its long-tongued moth, for example, the destruction of Madagascar's rainforest will certainly cause the extinction of the orchid. Since the moth is biologically tied to it, its extinction would be equally certain.

Pollination is by no means the only relationship between organisms where mutual interdependence has evolved, and not all relationships between species are beneficial ones. In general, biologists recognize three categories of conditions where organisms live intimately together or **symbioses**: **parasitism, commensalism**, and **mutualism**.[3] Many symbioses are clearly a result of coevolution; others not.

Parasitism

In the case of parasitism, the typical relationship is characterized by one symbiont, the parasite, benefiting from the relationship while the other, the host, is harmed. Some parasites require a specific or **definitive host** in which they must sexually reproduce. In the case of the protozoans that cause human malaria, that role is filled by certain species of the mosquito genus *Anopheles*. It is within humans, however, that malaria has a true parasitic relationship, as it is within their human hosts that malaria parasites increase in number, asexually, and cause harm. Human malaria parasites have probably coevolved with both *Anopheles* mosquitoes and humans, because no other mosquito genus can transmit them, let alone host them for sexual reproduction, and they appear to be incapable of parasitizing any vertebrate other than humans.[4] In contrast, the botfly, whose larva typically parasitizes a mammalian host, lives freely and reproduces as an adult, and it apparently can successfully parasitize any mammal it happens to find. As mentioned in an earlier chapter, the human botfly successfully parasitizes cattle in Costa Rica. Consequently, one would conclude that it has not undergone coevolution with any particular host.

Parasites such as malaria that live inside of a host are generally referred to as **endoparasites**, but **ectoparasites**, those that live on their hosts, exist as well. For example, ticks and lice are common ectoparasites in the tropics, as they are elsewhere. Mosquitoes and other biting insects are also considered to be parasites, although they do not technically live with their hosts. It can be argued that their lifestyle is closer to that of a predator in that they essentially stalk and attack rather than live in or on, but because they do not kill their prey in order to make a meal of it, the term predator simply does not apply.

Commensalism

Harvard biologist Edward O. Wilson argues that symbiosis increased biodiversity, and he is in all probability entirely correct. A living organism is a niche, perhaps several, that enterprising organisms can exploit. Gorillas at the Karisoke Research Centre in Rwanda, for example, harbored a variety of roundworms, tapeworms, protozoans, flukes, and more,[5] organisms that most likely would not exist if not for the gorillas. But by definition, a parasite harms its host, not necessarily killing it, but possibly weakening it to the point where some other parasite, a virus for example, may do the host in. This is not in the parasite's best interest, as it will die with the host. In the course of evolution, however, a parasite's impact can be mitigated. As an example, measles, now largely an irritating childhood disease was once often lethal. However, as susceptible individuals were eliminated from the population, those able to survive essentially reached an accommodation with the disease. Hypothetically, humans and measles viruses could coevolve to the point where the virus continues to use us for its reproduction but has no impact on our health; that is, it would benefit from its association with us, but we would be unaffected. By definition, it and we will have reached a commensalistic relationship.

Commensalistic relationships abound in the tropics. I would imagine that many involve internal commensals that began as parasites and changed as a result of coevolution, as may be the case with ourselves and measles, but others, while they may or may not have involved coevolution, certainly did not began with parasitism. For example, John Kricher describes the relationship between the sloth, an arboreal mammal, and a small moth. The moth lives in the sloth's fur and lays eggs in the sloth's dung. The moth benefits in having a home as well as a food supply and nursery for its larva, but there's no obvious benefit or harm to the sloth. Kricher argues that this does not represent coevolution because the moth has not influenced the evolution of the sloth.[6]

One of the more bizarre commensalistic relationships exists among orchids and their pollinators and involves a phenomenon called **sexual mimicry**. Usually, this means members of one sex of a particular species tend to resemble the other. In another case, the floral structures of some tropical, epiphytic orchids are arranged to resemble certain female bees or wasps. A particular species of orchid generally mimics a particular species of bee or wasp. In a number of cases, the orchid also mimics the sex-attractant **pheromone** of the female insect. A male bee that is attracted to the orchid attempts to copulate with the lure and, in the process, gets dusted with pollen. If that insect is lured to another floral decoy before finding a true female of its species, pollination of the orchid is completed.[7] Once again, this is technically not a symbiosis because the orchid and insect do not live together, and this is not a case of coevolution, as the orchid has not, in all likelihood, affected the evolution of the bee. However, the orchid has gained the advantage of an at least partially dedicated pollinator without having to invest in giving anything to the bee in return. The bee is not seriously harmed by the relationship other than in the loss of whatever energy is required by the attempted copulation. However, the orchid has evolved itself into a biological corner, as has the sloth's moth or any other specialized commensal. If anything were to happen to the independent organism, the survival of the commensal would be unlikely. In contrast, an unspecialized commensal like a bromeliad could be epiphytic on any tree. The extinction of one species would probably not affect it.

True commensalistic symbioses are clearly evident within the rainforest, however. Epiphytes that live on trees, insect and frog larvae that live in the tanks of bromeliads, and tiny arthropods that live on the lichens that are epiphytic on larger plants benefit from their hosts while causing no known harm—all are abundant throughout rainforests. The complexity of these associations and the abundance of them support Wilson's argument that symbiosis increases diversity.

Many relationships that appear to be commensalistic may in fact turn out to be something else, and many relationships now thought to be casual may turn out to be commensalistic with further study.

Mutualism

By definition, this is where two organisms live together for mutual benefit, and this is the relationship many people have in mind when they use the term symbiosis. Many mutualists have coevolved to the point where they are not only mutually beneficial, but mutually dependent as well. Moreover, they may impact many other organisms as well.

The classical example of mutualism is the lichen, an organism consisting of an alga and a fungus living together. These are extremely hardy and tenacious organisms, often living on the bark of trees but also living on bare rock in harsh environments where few other organisms could survive, and certainly where neither symbiont could survive alone. Lichens are often pioneer organisms, colonizing bare rock or soil denuded after a severe fire. As such, they begin the successional process that allows larger plants the opportunity to colonize the area, eventually allowing a disturbed area to recover. Indeed, it's questionable that forest ecosystems would have ever evolved were it not for the conditions generated by lichens. In many respects, mutualism can be described as the engine that drives much of the world's ecosystems, tropical forests among them.

One of the most important symbiotic relationships in that regard is nitrogen fixation. Nitrogen is one of the principal soil nutrients that plants need, but it is normally not abundant in rainforest soils. Consequently, rooted plants with symbiotic nitrogen-fixing bacteria, legumes, for example, are at an advantage over those that must obtain their nitrogen from soil sources. In the case of legumes, actually the plant family Fabaceae, the most abundant nitrogen-fixing plants, soil bacteria of the genus *Rhizobium* infect the plants' roots and form colonies, around which root tissue expands and forms nodules (Figure 9-1). The *Rhizobium* are able to take molecular nitrogen from the air and convert it to compounds that can be absorbed and used by the plant, which provides the bacteria with nutrients. Legumes are abundant in the American tropics and in Africa as well. Indeed, many of the standing trees, including emergents, are legumes, and they contribute very strongly to the diversity that exists there.[6] Moreover, many nonleguminous plants are able to take advantage

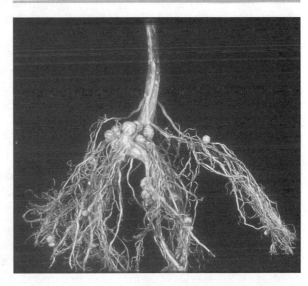

Figure 9-1 Legume roots showing *Rhizobium* nodules.

of fixed nitrogen that "spills over" from legumes. There are nitrogen-fixing bacteria that live freely in the soil, but they appear to be less efficient than *Rhizobium*. Symbiotic nitrogen fixation also occurs in the hindguts of termites[8] and among certain cyanobacteria (blue-green algae), lichens, and liverworts and in the mycorrhiza of diptocarps in Asian tropical forests, as mentioned earlier. The cyanobacteria may live mutualistically with other plants, as with the tropical water fern *Azolla*, or epiphytically on other plants. In either case, it can enrich its immediate environment with nitrogen, which other plants then exploit. Moreover, with the abundance of termites in tropical forests, they may contribute significantly to the tropical nitrogen cycle.

Mycorrhiza represent another case of mutualism that supports rainforest biodiversity. Discussed earlier, these are soil fungi that form associations with plant roots. The fungi absorb mineral nutrients from the soil, which they pass on to the plant. In return, the plant supplies the fungus with carbohydrates. According to Edward O. Wilson, it is questionable that plants could have ever evolved to the point they have and colonized land without mycorrhiza.[3]

Symbiotic nitrogen fixation and mycorrhiza are by no means restricted to tropical rainforests, but their significance in terms of biodiversity may be most appreciated there. The overwhelming bulk of rainforest biodiversity occurs in the canopy among the epiphytes, arthropods, and other arboreal organisms. New canopy species are frequently discovered when a tree falls. However, all of these organisms, discovered or otherwise, would not exist were it not for the standing trees that support them. Moreover, our own ancestors were tropical and arboreal. Were it not for the nitrogen fixation and mycorrhiza that allowed for the evolution of standing trees, we would not be here.

Mutualism is carried to an extreme when two species become so interconnected that they become totally mutually dependent. Examples of this are common enough, but one of the best is the fig and the fig wasp. In this case, the fig produces totally enclosed clusters of flowers, into which a pregnant species-specific fig wasp, covered with fig pollen, burrows. While searching for places for her eggs, she crawls over pistilate (female) flowers to which the pollen adheres, and pollination is accomplished and fruit and seeds develop, simultaneously with staminate (male) flowers and the wasp's eggs. After laying her eggs, she dies. Male offspring hatch first and immediately begin to search for females, whom they impregnate while still in the egg. The males then die, and the pregnant females hatch simultaneously with the maturation of the staminate flowers. As the females attempt to leave the fig, they crawl over the staminate flowers, get covered with pollen, and then, depending on the species of fig they are symbiotic with, either burrow out of the fruit or escape through a fruit opening. In either case, they then begin to search for a receptive flower without delay, as they live for only about a day. Once inside the flower, the female wasp lays her eggs and dies, but the cycle has been completed.[6] The relationship between the fig and the wasp must be an old one, because they have coevolved mutual dependence as well as mutual reliance. Attracting a random pollinator or finding a place to lay eggs is no problem; as long as one exists, the other can probably expect to. But again, bizarre twists seem to evolve as routinely as do symbioses.

In addition to the wasps that fertilize figs, there are wasps that parasitize them as well. These are as dependent upon the fig as is the pollinating wasp, but they do not pollinate; they only eat the fruit, thus harming the plant. However, some fig plants are also populated by ants that prey on the parasitic wasps, thus functioning as what a group of scientists who studied them referred to as "indirect mutualists."[9] If your head is starting to spin trying to keep track of who's symbiotic with whom, you're beginning to appreciate how diverse the rainforests are.

The indirect ant mutualists of fig trees are probably a case of ants taking advantage of an opportunity. Unless experiments showed some interdependence between the ant and the fig, it would be presumptuous to declare that coevolution has occurred or that a true symbiosis has developed. However, mutualistic symbioses between ants and plants are quite common in the tropical rainforests, and they occur in other ecosystems as well. Plants that harbor ants, known as myrmecophytes (Figure 9-2), initially attract their guests by providing food, in the form of nectar, shelter, or both. Most flowering plants that produce nectar do so in their flowers to attract pollinators. Ants are not typical pollinators, however, and the ant plants produce extrafloral nectaries, that is nectar sources outside of the flower. Some epiphytic myrmecophytes have aerial roots out of which ants are able to construct nests; others may have internal hollow areas where the ants live. For their part, the ants protect the plant. Most ants are well capable of biting, but some tropical ants sting or spray noxious chemicals as well, and most myrmecophytes ants do one of the latter as well as biting. The ants vary in their

Figure 9-2 Thorny acacia plant, a myrmecophyte.

aggressiveness, but some literally erupt out of the plant when it is disturbed and attack whatever is disturbing the plant *en masse*.[10]

The complexities of ant symbioses at times can be staggering. In addition to the mutualistic relationships with plants described above, there are associations with other animals. For example, some species of ants will tend aphids the way humans tend cows. In return, the ants obtain honeydew, a secretion of the aphids, much as humans obtain milk. Ants also have interspecific parasitic relationships where ants will raid other ant species' nests, capture larvae, and turn them into slaves. There are also ants and beetles that essentially fool myrmecophytes by mimicking the symbiotic ant and thus exploit the nectaries. Another perhaps unique ant symbiosis involves a three-member relationship. The leaf cutter ants, described in the previous chapter, cultivate a fungus upon which they feed. Neither the ant nor the fungus can live without the other. However, it has been recently discovered that a third symbiont, a bacterium, grows on the ants and produces an antibiotic that prevents the growth of pathogens within the fungal gardens. This three-part relationship may be the result of 10 million years of coevolution, but more significantly to us is the fact that the antibiotic(s) has remained effective. Many antibiotics used by humans have lost their effectiveness. The bacteria of the leafcutter colonies is a form a *Streptomyces*, a type from which some of our antibiotics have been isolated.[11] At this point, it would be premature to say that leafcutters' *Streptomyces* has potential benefit for ourselves, but it might. Nobody knows what other potential human benefits may result from learning about the organisms of the rainforest, but had the forests been destroyed and the leafcutters exterminated before this discovery was made, it would be impossible to even speculate on, let alone research, any potential it might have. In essence, preservation of the rainforest, its inhabitants, and their symbionts could very well benefit us as well.

A further complicated relationship is when an organism appears to be harming a symbiont it is actually helping. An example of this would be **frugivory**, the eating of fruit by animals. Fruits contain seeds, of course, and animals that eat fruit, if they're large enough, are likely to swallow the seeds. Seeds swallowed whole generally are not digested as one might expect. They pass through the animal's gut and are deposited at a later time when the animal defecates. In this way, the seed is removed from the parent plant; it does not have to compete with the larger, well-established parent for sunlight or soil nutrients. Moreover, it's beneficial to tropical animals to be able to eat fruit, as the year-round growing season pretty well guarantees that fruit is always available. Consequently, more than half of all species of tropical trees and shrubs depend on frugivores to distribute their seeds, and many of the vertebrates in tropical rainforests are frugivores. In general, the American tropics show a greater diversity of frugivores than do those of Africa and Asia.[12]

Species of mammals, birds, and lizards eat fruit. Even fish get into the act. In the igapo forests, the Brazilian tambaqui, a distant relative of piranha, is important in dispersing rubber trees among other plants. It eats the fruit but does not digest the seeds.[6] In Costa Rica, the machaca consumes the fruits of riparian figs that fall into the rivers. The seeds that pass through the fish's gut do not germinate as quickly as those that were not eaten, but the fish tends to move the seeds upstream, something that would not otherwise happen. Consequently, it would appear that it's worthwhile to the fig to sacrifice some germination for greater distribution.[13]

One final example of coevolution and mutualism involves mammals, specifically bats. Northern hemisphere bats characteristically eat insects, see poorly, and rely on their sense of hearing to locate their prey. Such animals exist in the tropics as well, but so do nectar-feeding bats, which see rather well, especially at night, have a well-developed sense of smell, but cannot echolocate as do the insectivorous bats. The flowers upon which the nectar-feeding bats rely are generally white in color, probably making them easier to see at night, and have what has been described as a "batlike" odor (Figure 9-3). The

Figure 9-3 Bat-pollinated flower.

flower also produce a more proteinacious pollen that contains amino acids bats specifically need. The flowers are also deep so that the bat's face gets covered with pollen as it feeds. **Chiropterophilous**, or bat-loving, plants and bats again demonstrate that coevolution generally leads to mutual dependence. The ultimate cost is that the loss of one guarantees the loss of the other.

In Closing

Interrelationships between organisms occur more abundantly in tropical rainforests than in most, if not all, other ecosystems as a result of and because of the greater diversity of life the rainforests contain. These relationships may be true symbioses with the participating organisms living together, as with the leaf-cutter ant and its fungus, or it may be more casual, as with a plant and a pollinator. The relationships benefit at least one of the participants but may or may not benefit the other, as in mutualism and commensalisms respectively. In the case of parasitism, one of the participants is actually harmed. Relationships may come about as a result of both participants making evolutionary adjustments to each other, coevolution, or only one doing so. In the latter case, both participants may be mutually dependent on one another for survival as well. In any event, the abundance of interrelationships demonstrates the complexity of life in the tropical rainforests.

References

1. Angier, N. 1992. It may be elusive, but moth with 15-inch tongue should be out there. *New York Times*. http://query.nytimes.com/gst/fullpage.html?sec=health& res = 9E0CE7DF163CF937A25752C0A964958260.
2. Saxer, I., and A. L. Rosenberger. 2005. Adaptations: The case of the penguins. *Visionlearning*. Vol. BIO-2, No. 6. http://www.visionlearning.com/library/module_ viewer.php?c3=1&mid=68&l=68.
3. Wilson, E. O. 1992. *The diversity of life*. Cambridge, MA: Belknap Press.
4. Marcus, B. A. 2004. *Deadly diseases and epidemics: Malaria*. Philadelphia: Chelsea House.
5. Redmond, I. 1983. Summary of parasitology research, November 1976 to April 1978. In Fossey, D. *Gorillas in the Mist*. Boston: Houghton Mifflin.
6. Kricher, J. 1997. *A neotropical companion*. Princeton, NJ: Princeton University Press.
7. Singer, R. B. A. Flach, S. Koehler, A. J. Marsaioli and M. D. C. E. Amaral. 2004. Sexual mimicry in *Mormolyca ringens* (Lindl.) Schltr. (Orchidaceae: Maxillariinae). *Annals of Botany* 93: 755–762. http://aob.oxfordjournals.org/cgi/content/full/93/ 6/755.
8. Lilburn, T. G., K. S. Kim, N. E. Ostrom, K. R. Byzek, J. R. Leadbetter, and J. A. Breznak. 2001. Nitrogen fixation by symbiotic and free-living sprirochetes. *Science* 292(5526): 2495–2498.

9. Schatz, B., M. Proffit, B. V. Rakhi, R. M. Borges, and M. Hossaert-McKey. 2006. Complex interactions on fig trees: Ants capturing parasitic wasps as possible indirect mutualists of the fig-fig wasp interaction. *Oikos* 113(2):344–352.

10. Schmidt, R. J. 1985. The super-nettles: A dermatologist's guide to ants in the plants. *International Journal of Dermatology* 24(4):204–210. http://bodd.cf.ac.uk/BotDerm Reviews/Myrmecophytes.html.

11. Dimijian, G. G. 2000. Evolving together: The biology of symbiosis, part 2. *Baylor University Medical Center Proceedings* 13(4):381–390. http://www.pubmedcentral. nih.gov/articlerender.fcgi?artid = 1312238#id2768614.

12. Fleming, T. H., R. Breitwisch, and G. H. Whitesides. 1987. Patterns of tropical vertebrate frugivores diversity. *Annual Review of Ecology and Systematics* 18:91–109.

13. Horn, M. H. 1997. Evidence for dispersal of fig seeds by the fruit-eating characid fish *Brycon guatemalensis* Regan in a Costa Rican tropical rain forest. *Oecologia* 1909(2):259–264.

Rainforest Resources: Foods, Medicines, and More

10

ASK SOMEBODY AT RANDOM FOR good reasons to protect the rainforests and you're likely to hear about potential life-saving medicines or sequestering carbon dioxide to prevent global warming. I couldn't dispute either one. In the case of medicines in particular, nobody has any idea how many potentially life-saving drugs may exist among the plants *and* animals that live there, and with increasing drug resistance on the part of microbes along with emerging diseases becoming more and more problematic, new medicines are going to be sorely needed in the future. However, I have another, more visceral reason for saving them. Some of the best things I have ever eaten come from, or originally came from, tropical forests. Many of these, of course, have been cultivated and are now grown commercially. But it is important to maintain the forests if for no other reason than to maintain a

111

genetic reservoir. Cultivated plants, especially when they are grown as crops, are subject to any number of pests and illnesses. Wild ancestors have often been found to have resistances that can be bred into their cultivated descendents. Moreover, there may even be as yet undiscovered rainforest foods as well as medicines, and it would be tragic to exterminate them before they can benefit us.

The number of resources in the tropics is immense; it is, once again, impossible to cover all of them in a book of this nature. Consequently, we will again discuss some representative examples for illustrative purposes. We'll start with food, work our way through medicines, and conclude with industrial products.

Foods

We're going to define foods broadly to include not only materials that we eat for energy and nutrients, but also those used for flavoring and, in some cases, preservation. The definition is arbitrary, but it allows for the coverage of a greater variety of rainforest products.

Chocolate

Retrieved from the cacao tree, a plant whose Latin name means food of the gods, if chocolate is not the most appreciated of all gifts bequeathed on our species by the rainforest, *it should be!* Its exact origin in tropical America is unknown, but it has been used by the Mayans, Aztecs, and other Central and South American native people for both food and medicine, and some cultures continue to do so. How the original exploiters learned to extract it from the bitter fruit and seeds of the cacao tree is a mystery. The Spanish explorer, Hernando Cortés recognized the value it would have in Europe. The eventual cultivation of cacao apparently was his idea.[1] The role of chocolate's taste in its popularity cannot be minimized, but both the Native Americans and Europeans believed it had medical value as well. More recently, research has shown that the medicinal value may be more than wishful thinking. Compounds in dark chocolate in particular have been found to function in reducing blood pressure and to increase beneficial chemicals in the blood when the chocolate is eaten in small amounts. White chocolate did not have the same effect.[2] Moreover, doesn't everyone know that a bit of chocolate can really give one's spirits a lift?

Unfortunately, the popularity of chocolate and the cultivation of cacao trees may prove to be its undoing. Originally from tropical America, most chocolate is now grown in Africa. Cacao trees in Central America are blighted;

commercial production of chocolate in Brazil declined by 75% between 1990 and 2000. Several fungal infections now plague cacao around the world.[3] During my last visit to Costa Rica, I was unable to purchase locally made chocolate, and indeed, the world would be a less pleasant place if chocolate were not cheaply and easily available. Unfortunately, cultivation often makes plants easier targets for pathogens in general, and globalization makes it easier for pathogens to find their targets.

Vanilla

Although it is not as universally extolled as chocolate, vanilla is another commercially important rainforest product. It is used extensively as a flavoring in many foods as well as being the primary flavor in confections. It grows worldwide in the tropics, but it was used originally in tropical America, Mexico actually, and it was introduced into Europe by Spanish explorers, as was chocolate. The flavoring is extracted from the pod of the vanilla plant, an orchid, that grows as a vine and, in Mexico anyway, has a designated pollinator, the melipona bee. There are a variety of species of vanilla, and today's commercial vanilla is a hybrid of the Mexican and others.[4]

Bananas

Possibly the most popular fruit in the world, bananas probably originated in Southeast Asia and have been cultivated for centuries. They are said to have been brought to Europe by Alexander the Great, and they have since spread around the world, being cultivated in tropics the world over. Bananas are internationally traded more than any other fruit, although only a fifth of total production is actually exported. The remainder is consumed in the country in which it is grown.[5] The yellow fruit that is typically eaten in the United States represents only one variety of a very diverse fruit. Indeed, much banana cultivation goes into the production of plantain, a starchy form of the fruit that is eaten as a staple in Africa. Occasionally one sees red bananas in grocery stores, but I'm told that there are blue and purple bananas and probably others as well.

The bananas we eat in the United States are largely grown in Central America. In Costa Rica, there is a supposedly love–hate relationship with the fruit. It is a major export and accounts for much of the country's economy, but it is grown on vast plantations that have been carved out of the rainforest. Moreover, it is grown in a **monoculture**, where it is the only plant growing (Figure 10-1). As described earlier, such conditions encourage pests, which must be kept at bay with massive amounts of pesticides that can

Figure 10-1 Banana plantation.

migrate into neighboring fields, forests, and waterways. Part of the high pesticide use may be due to US demands for cosmetically perfect fruit, (what a Central American friend of mine derisively refers to as "Barbie" bananas). Even so, much of the fruit is blemished and must be somehow discarded. Growers also use plastic bags to cover developing fruit to protect it from insect pests. These bags tend to migrate too; the wind often blows them from the farms into the rivers. Eventually, plantation soils become contaminated with agricultural chemicals and must be abandoned.[6]

Fruits in General

Among the pleasures one can experience in the tropics, enjoying fresh fruit almost right off the tree has to rank near the top. Very little compares to picking and eating a ripe orange off a tree or opening and enjoying a freshly fallen coconut. While no other tropical fruit has the economic impact of the banana, several others are still imported into the United States routinely. Oranges, for example, are grown in Florida and California, but imports of orange juice from Brazil and Mexico in September of 2005, for example, amounted to over $17 million, with the lion's share coming from Brazil.[7] While oranges and other citrus fruits can be produced in the United States, housing and commercial development on what used to be orange groves and our growing population have forced us to rely more and more on imports to satisfy our demands. Moreover, the climates of tropical countries make them more reliable for citrus production in general and orange production in particular. Brazil has routinely been the world's leading citrus producer,[8] a far distance from where the fruits originated in China.

Pineapples, originally from Brazil, are another tropical fruit that is routinely imported into the United States. Thailand and the Philippines are the two principal producers, while US production, restricted to Hawaii, has been declining.[9] Pineapples are one of the few members of the bromeliad plant family that are not epiphytes. They are low shrubs that are cultivated in long rows.

Other tropical fruits routinely consumed in the United States and available at grocery stores include mangos, papayas, and coconuts. Unless you live in Hawaii, it is probable that any of these that you enjoy are imported. In the case of mangoes and papayas, they probably came from Brazil or Mexico. Coconuts are used to manufacture other products in addition to food. Indonesia, India, and the Philippines are major producers. The papaya originated in tropical America. Mangoes and coconuts probably came from tropical Asia. There are other fruits as well, and countries other than the major exporters where they grow. Some may be grown strictly for local consumption. As humans have moved around the planet, they have taken many of their favorite foods with them and have often succeeded in transplanting them to their new homes, providing that the climate was agreeable. Plants that were originally from the rainforests will need rainforest climates In order to grow. Consequently, rainforest destruction and its ensuing local climate change may have a serious impact on our enjoyment of these products.

Tropical Nuts

Undoubtedly there are many members of this category, but three typically come to mind: macadamias, cashews, and Brazil nuts.

Most Americans typically think of macadamia nuts being as being a product of Hawaii, and indeed, over 20,000 tons a year are routinely produced there. However, the nut originated in the tropical rainforest of Australia, and that country continues to be the world's leading producer. Kenya in Africa and Guatemala, Brazil, and Costa Rica in the American tropics are also major producers.[10, 11]

Brazil nuts originated in the rainforests of South America, of course, where they continue to be principally found. Some production of Brazil nuts occurs on plantations, but for the most part, the trees grow wild, and local people gather the nuts with little or no effect on rainforest ecology.

Consequently, this would seem like an ideal argument for rainforest conservation, where preserving the forest would continue to provide a living for people living within it, particularly when many of the same people extract natural rubber when Brazil nuts are unavailable. Unfortunately, increasing human populations in the regions where commercial gathering of nuts now occurs are likely to put more pressure on the standing crop than it can support, and immediate economic demands may overwhelm longer-term conservation considerations.[12]

Cashew nuts are also relatively environmentally benign. They are generally not grown on plantations. Rather, they are grown as secondary crops on small farms or gathered wild. Much of the cashew tree contains toxic chemicals, which discourage animals from eating it. Moreover, the toxins in decaying leaves discourage competitive plant growth. Consequently, there is no overwhelming need for pesticides. Originally from northeastern Brazil, cashews generally grow in single species stands, and they are quite tolerant of dry conditions. As with many other rainforest plants, they have been transplanted to other parts of the world, and now several African countries and India, as well as Brazil, export the nuts.[13]

Medicines

In the motion picture *Medicine Man*, the principal character, a botanist played by Sean Connery, found a cure for cancer and then lost it. Much of the movie dealt with his finding the cure again before the bulldozers arrived to level the section of rainforest in which he was working. The movie played on the popular belief that the cures for all human ills are somewhere in the rainforest waiting to be discovered. Whether or not the belief is correct will probably never be determined, but it cannot be disputed that cures, or at least treatments, for some diseases exist there. Ethnobotanist Mark Plotkin has documented numerous instances of South American **shamen** using potions made from rainforest plants to treat illnesses, including sickle-cell disease.[14] As pointed out earlier, rainforest plants often contain naturally toxic compounds that are protective against herbivory by insects and other animals. In small doses, many of these compounds have medicinal value, and humans are not the only animals to have seemed to have figured that out. Chimpanzees have been observed to eat fig sap in Tanzania to purge themselves of roundworm parasites, and sick chimps have been seen eating plants that contain antibiotic compounds while healthy animals avoided them.[15] Not all natural cures come from plants nor from the rainforest, but in the early 1990s, roughly one quarter of all prescription drugs sold in the United States contained plant products—half of those were from the tropics. The value of those compounds was greater than $6 billion per year.[14] As with foods, animals, and other subjects, the list of important medical compounds that have been isolated from rainforests is long, and only a sampling will be discussed here.

Rosy Periwinkle

Found on the island of Madagascar, the rosy periwinkle (Figure 10-2) is a distant relative of magnolias, and it is now endangered in the wild owing to habitat destruction, although cultivated varieties are grown in much of the tropics and elsewhere. Two potent anticancer compounds have been isolated from this plant, and two more have been synthesized from one of the compounds. The compounds prevent cells from dividing, which is important in slowing or halting tumor growth. The plant has a long history in folk medicine, having been used to treat a variety of ailments, including diabetes.[15]

Cinchona Trees

Cinchona trees are native to South America, where Indians have long used extracts from them to treat fever and leg cramps. It is now known that the

Figure 10-2 Rosy periwinkle.

effective compound in cinchona is **quinine**, which has turned out to be the first effective treatment for malaria. Technically, quinine does not cure malaria, because it does not kill all forms of the parasite at all stages of its development, but when malaria was introduced into South America by European explorers, the Indians turned to cinchona extract to treat the fevers, and it turned out to cure or nearly cure the malaria in many cases. Today it remains the most effective treatment in some parts of the world. The demand for quinine almost led to the extinction of cinchonas by the mid 1800s. However, the trees were introduced for cultivation into Asia. The extinction was avoided, but not for conservationist reasons.[16,17]

Additional Medications

As noted above, medications derived from the rainforest are numerous, although the plants from which they are derived might not be as well known as the two described above. The drug reserpine, for example, used to treat high blood pressure, comes from the understory plant *Rauwolfia serpentina*, native to India, Sri Lanka, Burma, Myanmar, and Thailand. I believe that it is not strictly a rainforest plant, but it is evergreen and tropical. It has been used in native cultures to treat insect bites, poisonous snake bites, and fevers.[18]

Curare, a vine, has long been used by Amazon Indians to prepare poisons for the tips of their blow darts. The roots and stems of the vines are crushed and cooked, along with additions from other plants and animals. The darts are then dipped in the preparation and used to hunt game. Blowguns are made from bamboo, and the dart is launched at a prey animal. If it hits its target, usually an arboreal animal, the poison brings about a relaxation of the animals muscles until it falls from the tree. Medically, the Indians used the plant to treat urinary and menstrual problems. In Western medicine, it has been used as a surgical anesthetic and muscle relaxant.[19]

Poison dart frogs produce skin secretions that have also been used by Indians on their blowgun darts. The variety of compounds in these secretions is broad, and they have been found to have possible medical benefits in the treatment of pain and neurological disorders, including Alzheimer's disease.[20]

Construction and Industrial Materials

Most discussions on rainforest conservation seem to center around conserving biodiversity and preventing global warming, as mentioned earlier. However, there is a reasonable argument for the use of rainforest products, and there are proponents for **sustainable** exploitation of these products. Once again, the examples provided represent only a fraction of the material that comes out of the rainforests.

Timber

It would seem to be only a matter of common sense that timber would be a product of forests. After all, the most obvious component of a forest is trees. In the rainforests, tree species such as mahogany (Figure 10-3) and teak are large trees that provide wood that is highly valued in North America and Europe. Unfortunately, many of the trees are cut down illegally and are not replaced. In addition, cutting roads into rainforests and operating heavy equipment destroys more of the forest and causes soil erosion. Moreover, once the roads have been cut, it's easier for settlers to move in, cut down more rainforest, and establish farms.[21]

Kapok

Kapok is one of the largest trees in the rainforest, growing to a height of nearly 150 feet (50 m). It is one of the species described as an emergent. Although it is used for timber, its seeds are surround by a fibrous material that is used in stuffing for pillows, mattresses, and the like. It is also very buoyant and is used for the filling in floatation devices. Because new fiber is produced every year, it can be harvested without harming the tree, therefore providing at least a theoretically sustainable resource.[22] The issue is whether or not the economics of sustainable fiber production

Figure 10-3 Tree with lianas.

warrants keeping the trees standing over the short-term profits that would be made from cutting them down for timber.

Rubber

In many respects, rubber is much like the crude oil from which much of it is made. It is used world over by people who don't give it much thought. Originally a natural product from the rainforests of Brazil, about 70% is now made synthetically from petroleum, which is currently getting more and more expensive and eventually is going to become scarce. The bulk of natural rubber, which in some ways is superior to synthetic, and is at least in theory sustainable, is made from sap tapped from trees grown on plantations in Southeast Asia. Rubber trees in Brazil, where they originated, are threatened by a fungus that limits production and has largely caused the failure of rubber plantations there. Current research, however, seems to be getting the better of the fungus, as scientists have used selective breeding to develop resistant trees.[23]

In a number of ways, the story of Brazilian rubber encapsulates the struggles that have been going on in preserving the rainforests, particularly in Latin America. For many years, rubber trees have been tapped by people who lived in the remote villages in the forest. Rubber is made from the sap of the trees, which are able to withstand being tapped for years. However, much of the Brazilian rainforest destruction has been for the development of cattle ranching, and conflict between the ranchers and the rubber tappers has been ongoing. In 1988, Chico Mendes, a leader of the Brazilian rubber tappers union was murdered. Mendes had fought hard against the destruction of the rainforest.[24] While he may not have been entirely beloved by environmentalists who wanted the virgin rainforests left untouched, he argued that rubber production, even on plantations, was preferable to and more protective of the forests than cattle.[23]

Palm Oil

As with rubber, renewable energy would seem to be a desirable thing, and energy from **biofuels** should be renewable. Moreover, since biofuels are retrieved from plants that take carbon dioxide out of the air, using them for energy, in theory, should not contribute to global warming. In many respects palm oil appeared to be ideal, in that it burns cleanly and is the most energy rich of all plant oils. It is used in Europe both as a diesel fuel and for generating electricity. However, biofuels in general are turning out to be something less than they were promised in terms of environmental safety, and palm oil in particular has been especially problematic. To meet demands, palm oil is produced by farming, and much rainforest, particularly in Indonesia and Malaysia, has been destroyed in order to clear land for palm plantations. Moreover, many of these plantations have been located in former peat forests, and the draining and burning of peat land has actually contributed more carbon dioxide to the environment than using palm oil has saved. Further argument has it that if palm oil is produced in an environmentally friendly manner, it becomes very expensive.[25]

Biofuels in general are turning out to be somewhat disappointing in that their benefits do not come without consequences. Processing plants to produce energy requires energy, and growing fuel-yielding plants on rainforest soils often demands inputs of fertilizers, themselves often made from fossil fuels. The problem may not lie in the technology but rather in the volume of energy that is actually used, and an expanding human population with ever-growing demands can only mean that sooner or later, demand will outgrow supply. One would hope that a solution will be found before the forests are completely sacrificed. Palm oil is also used in the food processing industry where it is a component of many processed foods such as chocolate and snack foods. It has come under criticism for that because it is highly saturated, and saturated oils have been linked to heart disease.

Chiclé

On a bit of whimsy, we'll end this chapter by talking about chewing gum, which got its start with the Mayans and which has become a staple around the world. The historical basis of chewing gum is chiclé, (Figure 10-4) the latexlike sap of a Central American tree. Mayans used to chew the sap, a pastime that was picked up by Mexicans, including Antonio Lopez de Santa Anna, who became president of Mexico but is probably better known for the victory by his troops at the battle of the Alamo. A number of misadventures found Santa Anna in New York in 1886, stranded, but with some chiclé latex in his possession, which he hoped to persuade an inventor to turn into rubber. Thomas Adams consented to try, but he failed. However, with some experimenting, he came up with chiclé-based chewing gum.[26]

The chiclé tree grows wild and is tapped by Central Americans for its latex. Chiclé tappers slash chevron-shaped gashes in the trunk and collect the latex that drains out. Most chewing gum, as I understand it, no longer uses chiclé latex as a base. Rather, it uses an artificial material. However, when I was last in Belize, I learned that there is a growing market for natural chewing gum, and chiclé-based chewing gum is preferred to synthetic in Japan. Consequently, there is still money to be made by chiclé tappers. Practically

Figure 10-4 Chiclé tree.

every chiclé tree I saw in Belize had been tapped; their trunks were scarred where they had been gashed. When I finally saw one that had been left alone, I asked my guide if there was any particular reason for it. He looked at it for a couple of seconds and then acknowledged that he had never realized it was a chiclé tree. He had assumed that all had been tapped and so never bothered to learn what the tree looked like. He simply looked for scars.

In Closing

Once again, this chapter has considered only some representative rainforest products. In addition to commercial and conventional medicines, there are any number of herbal and cosmetic products one can buy. Cocoa and shea butter soaps and lotions are examples. Enterprising Costa Ricans are making paper out of banana leaves, and more than one Costa Rican has told me about industrial research being done on spider silk for products such as fishing line, bullet-proof vests, and parachutes. Undoubtedly there is much that can be done with rainforest products given appropriate manufacture, marketing, and ingenuity. The biggest single limiting factor is time. Will the rainforests last long enough for inventors and entrepreneurs to do the work to make the products available?

References

1. Dillinger, T. L., P. Barriga, S. Escárcega, M. Jimenez, D. S. Lowe, and L. E. Grivetti. 2000. Food of the gods: Cure for humanity? A cultural history of the medicinal and ritual use of chocolate. *Journal of Nutrition* 130:2057S-2072S. http://jn.nutrition. org/cgi/content/full/130/8/2057S#SEC5.
2. Taubert, D., R. Roesen, C. Lehmann, N. Jung, and E. Schömig. 2007. Effects of low habitual cocoa intake on blood pressure and bioactive nitric oxide: A randomized controlled trial. *Journal of the American Medical Association* 298(1):49–60.
3. Bowers, J. H., B. A. Bailey, P. K. Hebbar, S. Sanogo, and R. D. Lumsden. 2001. The impact of plant diseases on world chocolate production. *Plant Health Progress*. http://www.apsnet.org/online/feature/cacao/.
4. Evans, M. 2006. Vanilla odyssey. *Gastronomica: The Journal of Food* 6(2):91–93.
5. InfoCom: M@arket Inform@tion in the Commodities Area. 2007. *Information on banana*. United Nations Conference on Trade and Development. http://www. unctad.org/infocomm/anglais/banana/sitemap.htm#site.
6. McCracken, C. 1998. The impacts of banana plantation development in Central America. *Tripod*. http://members.tripod.com/foro_emaus/BanPlantsCA.htm.
7. Florida Department of Citrus. 2006. U.S. orange-juice imports. *Economic & Market Research Report No. IM-06-9*. November 14, 2006. http://www.floridajuice.com/ user_upload/files/ojim0906_455a2dfc6a62f.pdf.
8. United States Department of Agriculture. 2005. Situation and outlook for citrus. *World Horticultural Trade & U.S. Export Opportunities*, April 2005. http://www.fas. usda.gov/ htp/Hort_Circular/2005/04-05/04-08-05%20Citrus%20Feature.pdf.
9. Economic Research Service, USDA. 2003. Commodity highlight: Pineapple production concentrated in tropical regions of the world. *Fruit and Tree Nuts Outlook/ FTS-307/Nov. 21, 2003*. http://www.ers.usda.gov/Briefing/FruitAndTreeNuts/ fruitnutpdf/pineapple.pdf.
10. USDA, Foreign Agriculture Service. 2002. Situation and outlook for macadamias. *FASonline*. https://www.fas.usda.gov/htp2/circular/2000/00-03/maca.htm.

11. California Rare Fruit Growers, Inc. 1997. Macadamia. *CRFG Fruit Facts.* http://www.crfg.org/pubs/ff/macadamia.html.

12. Mori, S. I. 1992. The Brazil nut industry—past, present, and future. *The New York Botanical Garden,* http://www.nybg.org/bsci/braznut/. Reprinted from Plotkin, M. and L. Famolare (eds.) *Sustainable Harvest and Marketing of Rain Forest Products.* Washington DC: Island Press.

13. World Wildlife Fund. 2005. Agriculture and environment: Commodities: Cashews (*Anacardium occidentale*). http://www.panda.org/about_wwf/what_we_do/ policy/ agriculture_environment/commodities/cashews/environmental_impacts/index.cfm.

14. Plotkin, M. J. 1994. *Tales of a shaman's apprentice: An ethnobotanist searches for new medicines in the Amazon rain forest.* New York: Viking Press.

15. Plotkin, M. J. 2000. *Medicine quest.* New York: Viking Press.

16. Anon. 1996. Pharmacology of vinblastine, vincristine, vindesine, and vinorelbIne. Bio Tech Resources. http://biotech.icmb.utexas.edu/botany/vvv.html.

17. Marcus, B. A. 2004. *Deadly diseases and epidemics: Malaria.* Philadelphia: Chelsea House.

18. Anon. 2006. *Plant-derived drugs: Reserpine.* PhytoMedical Technologies, Inc. http://www.phytomedical.com/plant/reserpine.asp.

19. Anon. 1996–2005. *Tropical plant database: Curare.* Raintree Nutrition, Inc. http://www.rain-tree.com/curare.htm.

20. Mahidol, C., S. Ruchirawat, H. Prawat, S. Pisutjaroenpong, S. Engprasert, P. Chumsri, T. Tengchaisri, S. Sirisinha and P. Picha 1998. Biodiversity and natural product drug discovery. *Pure and Applied Chemistry* 70(11):2065-2072. http://www.iupac.org/ publications/pac/1998/pdf/7011x2065.pdf.

21. The Rain forest Foundation. 2004–2007. *Rain forest timber.* http://www.rain forest foundationuk.org/s-Rain forest%20Timber.

22. Kapok tree. *Blue planet biomes.* http://www.blueplanetbiomes.org/kapok.htm.

23. Couper, H., and N. Henbest. 2007. Green gold: How a Brazilian forest of rubber trees is bouncing back. *The Independent* September 10. http://news.independent.co. uk/sci_tech/article2595355.ece.

24. Comitê Chico Mendes. http://www.chicomendes.org/index.htm.

25. Rosenthal, E. 2007. Once a dream fuel, palm oil may be an eco-nightmare. *New York Times* January 31, 2007. http://www.nytimes.com/2007/01/31/business/ worldbusiness/31biofuel.html?pagewanted=1&ex=1327899600en= e653a375e67e8e49ei=5088partner=rssnytemc=rss.

26. Heron, S. 1999. *The economic botany of Manilkara zapota (L.) Van Royen.* Southern Illinois University Carbondale/Ethnobotanical leaflets. http://www.siu.edu/~ ebl/leaflets/zapota.htm.

Rainforest Destruction

11

IN MY EXPERIENCE, RAINFOREST destruction, deforestation actually, is perceived to be the second gravest threat facing the planet: global warming is first. Of course, most of my experience has been with students who may have been saying what they thought the professor wanted to hear. At least they'd heard of the problems, and in the minds of many people, the two are inexorably linked. Possibly, but the roll of the rainforest in global warming is more appropriately discussed in Chapter 12. In previous chapters, we briefly and intermittently discussed some consequences of rainforest destruction. In this chapter, we'll look at it in more detail.

The Scope of the Problem

Anyone looking for information on how much of the Earth was originally covered by tropical rainforest is going to find a variety of estimates. One that I seemed to come across most frequently suggested that it may have been as much as 15% as recently as 1950. Similarly, estimates over how much is left also vary, but a consistent figure seems to be around 7%. If these estimates are accurate, more than half of the total tropical rainforests on the planet have disappeared in a bit less than 60 years. At that rate, all will be gone by around 2050, another estimate one frequently sees. According to William Lawrence and colleagues, around 60% of the remaining rainforest is in the Amazon basin and covers around 2 million square miles (5.3 million sq. km).[1] Doing a little math, one would estimate that the total existing rainforest amounts to roughly 3.4 million square miles (8.8 million sq. km). In 1950, then, tropical rainforests covered approximately 7.3 million square miles (19 million sq. km). In essence, a lot of trees have been cut down. In all probability, many species have become extinct in the process, and whatever products or services could have been extracted from them will never be known. In addition to species loss, deforestation leads to soil erosion, which impacts waterways and coastal ecosystems. There are a number of reasons why deforestation is occurring so rapidly, and this chapter will focus on some of the major ones.

Logging

The most obvious rainforest resource to be exploited is the trees. Not all rainforest trees are desirable for timber, of course, but many that are, such as mahogany, are large and draped with vines and lianas that may pass to other trees within the canopy. Consequently, when such a tree is cut down, it not only knocks over other trees, it drags more down as well. Even trees that remain standing are often damaged by the removal of a large tree nearby. Moreover, epiphytes and arboreal animals in its crown will be destroyed, or at least disrupted. The result, then, is a large gap. Given time, the gap would grow in, but roads have to be constructed to get the cutting equipment into the forest and the logs out. Thus, even when selective cutting is practiced, the damage to the rainforest goes substantially beyond the loss of individual trees. Furthermore, the large trees generally are old trees; it often takes over a century for them to grow to harvestable size but less than an hour to be felled. As a result, entire populations of some tree species have been removed from some areas. In general, however, it is not in a logger's immediate best interest to practice selective cutting. Loggers (Figure 11-1) make their living selling trees, and the more they can remove from a small area in a short time, the higher their income. Unfortunately for the forests, there is no shortage of markets in wealthy countries in Europe, North America, and Asia for rainforest timber. Moreover, much of the logging now occurring is illegal.[1]

Figure 11-1 Rainforest logging operation.

In Africa and probably elsewhere, some deforestation is also occurring as poor people search for fuel. Wood is needed for cooking fires and the like, and standing timber is often the only source.

Part of the problem with logging is the cutting of roads. This not only damages plants, it also disrupts the soil and provides avenues for rain to run off, carrying soil with it. Thus, it exacerbates soil erosion. The roads also open the forests to hunters who shoot game to feed the logging crews. In addition, it allows access to the forest by landless farmers, who use the logging roads to move into the forest. They then clear whatever is left standing, often by burning, and begin small-scale farming operations.[2] Such activities are generally limited in time. Burning tropical soils causes a reduction in fertility within a few years, and the destruction of ground cover further aggravates erosion and the loss of nutrients.[3]

Agriculture

There are three different approaches to agriculture that appear to be occurring in the rainforests: slash and burn, cash crops, and cattle ranching. Each one affects the rainforest and contributes to deforestation differently, but in combination they account for massive amounts of rainforest destruction.

Slash and Burn

Also referred to as shifting cultivation, this has been suggested to be the biggest single agricultural threat to the rainforests. It is particularly prevalent in Africa, and it is largely a result of poverty and displacement. Landless peasants who for any reason have been displaced from land they had once occupied clear sections of rainforest by burning, as described earlier.[4] The burning releases minerals from the trees that temporarily enrich the soil, but the fertility declines within a few years, as does crop yield, and the settlers are forced to move on.

As mentioned above, the people who practice slash and burn have somehow been displaced from land they had previously occupied. It has been suggested that much displacement is a result of political and social actions, for example government or corporate clearing of land for agriculture, damming rivers for hydroelectricity, mining, and other commercial projects.[2] Political instability, war, and population pressures may also contribute to forcing people into the forests. Often they follow logging roads and settle in cleared areas.

Cash Crops

When I was an undergraduate, well before anyone was worried about rainforest destruction and even before the infertile nature of rainforest soils was well understood, it was not unusual to hear academics of various disciplines, including science, dismiss the then-beginning concerns about the growing human population and world hunger. They used platitudes about how the Amazon basin would become the breadbasket of the world. The idea persisted even when scientific studies revealed that this would never be the case. Much of the clearing of the Amazon for agricultural development had been encouraged by the government of Brazil.[5] Brazil has since enacted progressive laws to protect the rainforest, but they are difficult to enforce. The principal crop being grown in the Amazon rainforest is soy, and much of it goes to Europe to feed cattle and poultry that end up in fast-food hamburgers and chicken.[6]

The extensive planting of a single species of plant, a crop essentially, is known as **monoculture**. It was described in an earlier chapter with bananas. Monocultures may be food crops such as soybeans, but they may also be tree crops for lumber or oil as well as for fruit or nuts. In any event, all monocultures, even trees, have an environmental impact. They simplify ecosystems reducing biodiversity, and they invite specific pests. A caterpillar that feeds on banana leaves, for example, sees a banana plantation as a giant smorgasbord. Moreover, ecosystem simplification means that some of the organisms that are naturally a part of the ecosystem are eliminated, and some of these may be potentially valuable resources. In the case of trees, nonnative trees planted in plantations can have a serious impact on the understory plants.[7] Furthermore, whenever a crop is harvested, nutrients are removed, particularly nitrogen.[8] In the case of tree crops other than the trees themselves, the loss is small, but in the case of herbaceous and timber crops, the entire plant is harvested. Removal of non-cash-providing parts still removes nutrients from the soil. In the case of soybeans, a legume and host for nitrogen-fixing bacteria, the impact of nutrient removal may not be immediately evident. However, other minerals such as potassium and phosphorus are also removed, and unless they are replaced, soil depletion eventually results.

Cattle Ranching

Cattle have also proven to be a desirable product, and much rainforest devastation has occurred in order to create pastureland. In general, the cattle are not used locally but are exported, often ending up in fast-food hamburgers. Over 55% of Brazilian beef goes to the European Union and Russia. In Brazil, beef exports generate $3 billion a year.[5] Forests are cleared for ranching by burning, and cattle often graze almost to the root. In doing so, the cattle remove nutrients from the soil. These are completely lost when the cattle are removed from the pasture. The impact of cattle on the rainforest includes not only soil depletion but compaction as well. This reduces root and water penetration, as well as air circulation making it difficult for native plants to return. Moreover, as with plants, high concentrations of animals often invite pests or parasites, and bovine foot-and-mouth disease has become a problem in parts of Brazil.[9]

Mining

The destruction that results from large-scale mining is pretty easy to imagine. Huge scars on the landscape, soil erosion, and siltation of rivers as well as the destruction of trees come quickly to mind. Air pollution from dust and exhaust are additional consequences, and all of the fallout of road building discussed earlier exists as well. Large-scale mining operations include removal of bauxite for aluminum and other ores such as iron, copper, and nickel. But in the Amazon, even small-scale mining operations, particularly for gold, are causing serious environmental damage.

Intensive gold mining has been occurring in the South American rainforest since 1980, and it continues to occur. Much gold is found in **alluvial** deposits. It's separated from other materials in the sediments hydraulically and by amalgamation with mercury. Overflows carry mercury into streams, where it causes damage to fish and other organisms. In addition, sediments wash into streams as well, largely as a result of erosion. The sediments carry such metals as aluminum, iron, and potassium, but the suspended sediments are particularly damaging, causing increases in stream sediment load and siltation. Studies in Surinam have shown that streams affected by gold mining runoff experience reduction in fish diversity, including loss of young fish and food fishes. The problem appears to result from sediments reducing habitat diversity. As in other cases, much of the problem appears to be a consequence of small, often unregulated operations.[10]

Impacts of Deforestation

A question I often heard from students whom I took to Central America was some variant of "How can they allow the destruction of their rainforests?" The answer is complicated. For one thing, for most of history it was inconceivable to most people that anything as vast as the rainforests could be destroyed. Early warnings were often dismissed as alarmist. Moreover, destruction of the rainforest was often seen as economically beneficial. Many Latin American countries, for example, have large national debts. Rainforest clearing for agricultural development was willingly accepted with the idea of produce being sold abroad.[11] Selling timber abroad was also seen as a means of bringing cash into the country. As described above, it made for removal of more than simply the timber tree, and that often opened the door to people moving into the forest for slash-and-burn farming.

In addition to the soil erosion and consequent nutrient loss described earlier, forest clearing also disrupts the water cycle, which sets off a cascade of changes. A rainforest may recycle half of the rainfall that hits it.[1] Through the process of **transpiration**, water pulled out of the soil by trees evaporates back into the air, condenses, and precipitates again, actually forming a smaller cycle within the larger one. One can actually see this in progress as the water evaporated from trees is so concentrated, the vapor looks like smoke rising from the forests. Removal of the trees breaks the cycle, more water runs off carrying sediment and nutrients with it. Moreover, once the trees are gone, the soil loses protection from the sun. More moisture is evaporated from the soil, which eventually dehydrates.[3] The loss of transpiration reduces rainfall, thus exacerbating soil dehydration and also making remaining forests susceptible to fire. In general, fires are rare in old growth rainforest, other than in really dry **El Niño** years.[1] But large-scale tree removal changes the dynamics and introduces a greater probability of fire.

Part of the problem with deforestation is that it introduces what is described as the "edge effect" into the deep forest. Along forest edges, sunlight is able to penetrate often to ground level, sustaining more low plants and generally supporting more dense growth. Often, plants and animals that live in the deep forest are unable to survive along the edge. Furthermore, more desiccation occurs along the forest edges, drought becomes more of a problem, trees die, and the probability of fire increases even more.[12]

Popular belief has it that the rainforests are one of the Earth's primary defenses against global warming by sequestering the carbon dioxide they use in photosynthesis. Indeed, they do play a role, but it's a complicated one, and it is more appropriately discussed in the chapter on global climate change, later in the book. For the moment, suffice it to say that most mature forests tend to be carbon neutral: they emit as much carbon as they absorb.[13] Consequently, their role in global warming may be something different than what many people believe. However, rainforest clearing does increase carbon dioxide levels in the atmosphere, and that does exacerbate global warming.

One topic that most people rarely think about in terms of rainforest destruction is emerging tropical diseases. For example, SARS (severe acute respiratory syndrome), a type of viral pneumonia, apparently originated in Guangdong province in southern China and in Viet Nam.[14] In 2003, outbreaks of this disease occurred in Hong Kong, not far from its origin, and in Toronto, Canada, demonstrating that tropical diseases can be spread around the world. Another emerging disease that has world health officials worried is dengue fever, another viral disease that is spread by a particular species of mosquito, *Aedes aegytpti*. This species is able to reproduce in small ponds or any kind of container that will hold water, including discarded tires, birdbaths, puddles in streets, and discarded household debris one often finds around impoverished villages in developing countries (Figure 11-2). The incidence of dengue fever is increasing in Southeast Asia, apparently as population grows.[15] The mosquito, which also transmits yellow fever, is apparently highly adaptable; it has spread throughout the tropics and in the warmer, wetter states of the southern United States. Moreover, as more rainforests disappear and more people crowd into the lands formerly occupied by rainforest, their exposure to these diseases should increase. As the animals that harbor the parasites that cause the diseases become less abundant, the pathogens that cause them and the vectors that spread them will look for alternative hosts, and humans may be the most available.

Finally, the destruction of the rainforests also means the destruction of any indigenous cultures that remain within them. In the past, bringing so-called primitive societies into civilization (Figure 11-3) was considered to be desirable, particularly with regard to religion. Missionaries often accompanied explorers to spread Western religion as European countries spread their influence around the world. However, many indigenous people did not adapt

Figure 11-2 Village dump, where open containers can collect rainwater and provide breeding grounds for mosquitoes.

Figure 11-3 South American Indians in popular American dress.

well to European civilization, and today many former European colonies seethe with crowding and poverty, and as the lures of civilization attract aboriginal people, they often give up their culture. In South America, much of the native knowledge of rainforest foods and medicines is disappearing as the old **shamans** die off and few young people replace them. At this point it is questionable that cultures totally untouched by Western civilization still exist, but recent history with indigenous cultures has demonstrated that even simple contact can be harmful. Consequently, among some South Americans, it is believed that the best civilization can do for indigenous people is to leave them alone.[16]

In Closing

Standing rainforest now covers 50% or less of the Earth than it did 60 years ago. At current rates of deforestation, all of it will be gone by the year 2050. Destruction of the rainforests are occurring as a result of exploitation for timber and metals and in clearing for agriculture. Much of it is motivated by financial considerations, and the developed world continues to be the major market for rainforest products and produce. Thus, soybeans grown in Brazil feed cattle that end up as fast-food hamburgers in Europe and North America. The same can be said for tropically raised cattle. Rainforest destruction is also bringing about changes in the remaining, standing parts of the biome as edge effect, desiccation, and fires are becoming more problematic. Moreover, emerging tropical disease may now be potentially more capable of spreading to industrial countries as the reservoirs for these diseases disappear as human population grows and more people occupy the lands those reservoirs once inhabited. Disappearing rainforest is also a factor in global climate change.

Exactly how all of these changes will play themselves out is impossible to predict. However, it is probably safe to say that poverty will expand as resources dwindle, and the disappearing rainforest will at least coincide with physical and social changes that will make the planet a less pleasant place in which to live.

References

1. Lawrence, W. F., S, Bergen, M. A. Cochrane, P. M. Fearnside, P. Delamônica, S. A. D'Angelo, C. Barber, and T. Fernandes. 2005. The future of the Amazon. In: *Tropical rain forests: Past, present, and future, eds.* E. Bermingham et al. Chicago: University of Chicago Press.
2. Rain Forest Information Centre. Undated. The causes of rain forest destruction. *Rain Forest Information Centre* educational supplement. http://www.rain forestinfo. org.au/background/causes.htm.
3. Wood, M. 1995. *Environmental soil biology.* 2nd ed. New York: Springer.
4. South West Asia Project. 1999. *African rain forest project: Study of the Central African rain forests and their biodiversities using satellite imagery of diverse characteristics.* Center for Earth Observation, Yale University, and the International Institute of Tropical Agriculture (IITA), Africa; joint publication. http://www.yale.edu/ceo/ Projects/swap/Links/Africa/africa.html.
5. Wallace, S. 2007. Last of the Amazon. *National Geographic* 211(1):40–71.
6. Downie, A. 2007. Amazon harvest. *Nature Conservancy Magazine* 57(3):36–51.
7. Harrington, R. A., and J. J. Ewel. 1997. Invasibility of tree plantations by native and non-indigenous plant species in Hawaii. *Forest Ecology and Management* 99(1–2):153–162. http://www.sciencedirect.com/science?_ob=ArticleURL&_ udi=B6T6X-3ST7DP6-C&_user=10&_coverDate=12%2F31%2F1997&_rdoc= 1&_fmt=&_orig=search&_sort=d&view=c&_acct=C000050221&_version=1&_ urlVersion=0&_userid=10&md5=575145783843d348fc385906cf1dcea9.
8. Kellman, M., and R. Tackaberry. 1997. *Tropical environments.* New York: Routledge.
9. Project Amazonia: Threats—agriculture and cattle ranching. http://web.mit.edu/ 12.000/www/m2006/final/threats/threat_agg.html.
10. Mol, J. H., and P. E. Ouboter. 2004. Downstream effects of erosion from small-scale gold mining on the instream habitat and fish community of a small neotropical rain forest stream. *Conservation Biology* 18(1):201–214.
11. Vandermeer, J., and I. Perfecto. 1995. *Breakfast of biodiversity: The truth about rain forest destruction.* Oakland, CA: Food First Books.

12. Laurance W. F., and G. B. Williamson. 2001. Positive feedbacks among forest fragmentation, drought, and climate change in the Amazon. *Conservation Biology* 15(6):1529–1535.

13. Pregitzer, K. S. and E. S. Euskirchen. 2004. Carbon cycling and storage in world forests: Biome patterns related to forest age. *Global Change Biology* 10(2)2052–2077.

14. Xu R-H et al. 2004. Epidemiologic clues to SARS origin in China. *Emerging Infectious Diseases* June 2004. http://www.cdc.gov/ncidod/EID/vol10no6/03-0852.htm.

15. Anon. 2007. The prosperity bug. *Economist* 384(8537):46–47.

16. Angelo, C. 2007. Prime directive for the last Americans. *Scientific American* 296(5):40–41.

Rainforest Conservation and the Importance of Ecotourism

MOST AMERICANS ARE AWARE, or at least should be, that when Columbus discovered the Americas in 1492, he was actually looking for a more direct route from Spain to Asia than the overland trek that then existed. His voyages were fueled by competition for trade between Europe and Asia. It was popularly believed at the time that only open ocean existed between the west coast of Europe and the east coast of Asia, and Columbus mistakenly thought the Earth was smaller than it was. The idea that everyone at the time believed the world was flat is mythical, but many believed the distance across it was too great to be made by the small sailing ships of the time. His discovery of the Americas probably came as something of a shock, although it did give Spain a head start on colonizing the continent. Early Spanish explorers who followed Columbus to the Americas are said to have searched for

gold, but it was the timber of the New World that supported the shipbuild-
ing that made England ruler of the seas. When the Americas were discovered,
they were largely forested, and the impression most Europeans who saw
them probably had was that the resources were inexhaustible.

Anyone who suggested to a colonial European that the forests of the
New World, the rainforests in particular, could ever be exterminated would
probably have been ridiculed. Even 50 years ago, a doomsayer who predicted
the end of the rainforests would have been met with skepticism. The forests
of the Amazon alone were so vast. Moreover, the timber, wildlife, and other
resources available within the New World forests were a competitive goldmine,
figuratively speaking, for the European countries that colonized them, and the
economics of the time encouraged exploitation. Perhaps that might at least
partially explain why the coastal countries of Western Europe generally prospered in contrast to the landlocked ones in Eastern Europe. The development
of the United States has also occurred at the cost of the environment. The resources available on this continent were massive, and they were exploited
mercilessly. Water bodies were polluted with municipal and industrial sewage,
and soils were planted and trees harvested with no thought that they could
be exhausted. And if they were, there was always more farther west. For all
of human history, no thought had been given toward conservation; there never
seemed to be a need to. During colonial times it is probable that anyone who
predicted that the American passenger pigeon would become extinct would
have been ridiculed. Likewise, anyone in frontier days who had suggested
that the American bison would be brought to the edge of extinction would have
been received identically. Even today, with abundant evidence to the contrary, there are those who deny that any threat of annihilation of the rainforests
exists, or that if the rainforests were destroyed that it would have any dele-
terious effect on the planet or human population, in spite of scientific evidence
to the contrary. Human nature seems to make some people either incapable
or unwilling to admit that our species is capable of any kind of global destruction. History would suggest that they are mistaken.

A History of Destruction

UCLA Medical School physiologist Jared Diamond has documented numerous cases of past societies that collapsed because of overexploitation or other
misuses of their environmental support systems.[1,2,3] Such collapses have occurred all over the world, but a particularly pertinent example he describes
is the Polynesian culture that existed on Easter Island in the South Pacific. Remote and equally isolated from South America and other inhabited Pacific
islands, the Polynesians who settled on Easter Island were probably cut off

from all human contact once they discovered it. Archeological and paleo-ecological research suggests that they found a rich tropical forest, which they cleared for agriculture and wood, eventually deforesting the island. In so doing, they drove island bird populations to extinction and destroyed the trees they needed to make canoes for fishing. By the time Europeans got to Easter Island, the civilization that had flourished there and once supported a population of as many as 10,000 people had collapsed. Essentially a closed society without a highly developed technology, Easter Island can be thought of as a microcosm of planet Earth of today, where environmental resources are being exploited more rapidly than they can be replenished or replaced.[2] Not all island societies are doomed to collapse. Diamond also documents isolated Polynesian societies that have not shared the experience of Easter Island.

Ecosystem destruction may not necessarily be the immediate cause of collapse of civilizations, but it may be a contributor. In fact, it's rare that a single cause spells the end of a civilization. For example, there's much evidence to suggest that the Maya civilization in Central America, which may have had a population in excess of 13 million, collapsed around a millennium ago because of drought.[4] However, given a population of that size, it is likely that they were using resources faster than they could be replaced, and extensive deforestation is highly probable. The drought may have been the final issue.[3]

Today, much of the sentiment toward saving the rainforests is concentrated among people who are not directly affected by it. North Americans, Europeans, and even the economic elite in some rainforest countries can abstractly lament the destruction of the rainforests, but an impoverished farmer in the tropics can't afford that luxury. Moreover, it is unrealistic to expect someone who financially benefits from rainforest exploitation to support conservation. Ecological collapse is simply not his priority. Consequently, if the chain of ecological collapse is to be broken and the rainforests are to be conserved, it has to be in the economic best interest of everyone, or at least reasonable to everyone. Research is currently underway to find economic alternatives that suit as many as possible.[5]

Conservation

Perhaps a good place to begin this section is with a working definition of *conservation*. To a biologist, it is essentially the protection and maintenance of the world's biodiversity. In the minds of many, it is equivalent to preservation, which generally means leaving something untouched. By way of explanation, setting aside a section of rainforest in which hiking, camping, the harvesting of fruit off of trees, or even hunting would be consistent with conservation if the activities were **sustainable**, that is they were carried out in a manner that did not deplete or endanger any of the resident species. In the case of preservation, the section would be closed to all human activity.

At the moment, approximately 5% of the planet is set aside as parks.[6] Historically, this has been the typical way of conserving wild areas. Unfortunately, it's often not successful. In countries where there are high population densities and poverty, people are unlikely to respect rules protecting the parklands, particularly if there are such resources as food within them that the people need. In addition, poverty or even greed will motivate people to exploit protected resources as well. In other words, if people do not have some sound economic reason to participate in rainforest protection, they are not likely to.[7] In addition, there's the matter of mineral, timber, and energy interests whose profit motive for destroying the rainforest is much stronger than any interest they may have in conserving it. Moreover, while there may be emotional desires to maintain the rainforest on the part of the international community, there is also an international demand for rainforest products. When it comes to immediate economic gain, long-term conservation usually doesn't stand a chance.

One approach toward solving conflicting needs for conservation and exploitation is what has been called **sustainable** harvesting (Figure 12-1). Using timber as an example, sustainable harvesting would involve selective removal of trees at a rate that is no greater than new trees are added to the population, and so it should be able to be carried on forever. Theoretically, it's entirely possible. Vanclay and Preston ran a 500-year mathematical simulation where trees were harvested from wet rainforest in Queensland, Australia, with no loss of species or decrease in populations. Moreover, the authors argued that tree felling at sustainable rates contributes to biodiversity by creating gaps in the forest much as the toppling of a tree naturally would.[8] In practice it would present something of a challenge, because population densities of valuable trees in rainforests are low, and it would not offer much return to a lumberman to leave more timber trees standing than he harvests. Moreover, sustainable harvesting requires patience and restraint. Kammesheidt et al. determined that logged-over forests in Venezuela could be harvested sustainably with cutting cycles of not less than 60 years,[9] which is essentially once in a lifetime.

Fortunately, wood is not the only thing the rainforests produce. Fruits such as mangoes and papayas, seeds such as Brazil nuts and cashews, and plant exudates like rubber (latex) and chiclé, all discussed earlier, are examples of rainforest products that can be harvested without felling and killing the tree that produced them. Even structural parts such as the canes of rattan palms of the Asian rainforests can be harvested without permanently damaging the trees. However, harvesters have been known to cut down a

Figure 12-1 Tropical forest defoliated by timber extraction.

tree in order to get at its products, despite the fact that killing the tree will to-tally eliminate any future harvests.[10] Consequently, any program that pro-motes the sustainable harvest of nontimber components of the rainforests should probably include a strong education section.

Sustainable harvesting need not necessarily be restricted to gathering plant products. Cash crops like coffee, tea, citrus, bananas, and cocoa can be grown sustainably. One organization in particular, the Rainforest Alliance, certifies farms that produce sustainably,[11] and if I may insert a personal note, their certified coffee is good.

Oddly, the concept of sustainable agriculture in the rainforest is not that new. Archeologists have discovered areas in Brazil where the soil is dark, fertile, and full of carbon, unlike typical rainforest soils. Moreover, this soil ap-pears to have been made by pre-Columbian Indians rather than having formed naturally. Researchers are trying to learn how those ancient farmers accom-plished that feat, because improving fertility of and crop growth on soils cur-rently being farmed should relieve the need to destroy more rainforest in order to grow crops.[12]

Sustainable agriculture, if it can be accomplished, makes a lot of sense, as does organic farming or any other kind of conservation-oriented farming. Soil is the foundation of the agricultural pyramid, and conservation-oriented farming functions to protect and enhance the soil. In addition, conservation-oriented agriculture does not contribute to the migration of fertilizers, pesticides, and topsoil from farms to waterways. It is popularly believed that conservation-oriented farming cannot match the productivity of conventional farming, but the difference is minimal. Furthermore, in dry years, conservation-oriented farms actually outproduce conventional ones. Moreover, while studies on the issue are few, some show that fruits and vegetables from organic farms tend to be higher in nutrients and leave consumers with lower synthetic chemical residues than conventionally grown food.[13] Finally, during Hurricane Mitch in 1998, conventional farms in Honduras experienced 60–80% more soil erosion and damage than did conservation-oriented farms.[6] In support of the concept, Deere & Company, a leading manufacturer of agriculture equipment, is cooperating with a Brazilian agricultural company in restoring degraded pasture land for farming, rather than cutting down more forest. Land not suitable for farming is being reforested.[6]

Ecotourism

In an ideal world, the most attractive places on the planet would be the easiest in which to live. People would then not have to burn fuel and pollute the air in order to get to them. In addition, people would have little or no impact on their environments so that these attractive areas would be conserved naturally. No ecosystems would be destroyed. Unfortunately, this is not an ideal world and probably has never been one. Jared Diamond and others have written about human despoliation of the environment having occurred throughout history, although many of us have the sense that it's probably occurring more rapidly now than it has at any time in the past. If that is indeed the case, it is largely due to the immensity of our population and the advanced state of our technology. Still, many of us try to escape the crowding and the congestion by traveling to those attractive, lightly settled places including, but not restricted to, the tropical rainforests. Reasons why they're lightly settled, however, may include that they're often remote and are in some way or ways undesirable places to live. The human spirit being what it is, however, some rugged individuals manage to gain a foothold in those places, and others often follow them. Thus, cities like Phoenix now flourish in a desert while others like New Orleans occupy swamps. There are consequences to living in such places, such as severe heat and potential water shortages in the former and the risk of being flattened by hurricanes in the latter, but people still find the areas desirable. Unfortunately, the more people that settle in an area, the more ecological damage it is likely to sustain. Moreover, people do not necessarily have to settle to cause ecological damage. Enough people coming in and visiting can have a detrimental impact. For example, tourists trampling on the reefs in the Sulu-Sulawesi region of Tunku Abdul Rahman National Park in Malaysia are contributing to their destruction.[14] Still, people are going to travel to appealing destinations, indigenous people need to earn money, and fragile ecosystems have to be conserved. A partial solution would seem to be to encourage tourism that does no damage to the ecosystem. This is close to the International Ecotourism Society's definition of **sustainable travel**.[15] Travel "that conserves the environment and improves the well-being of local people" would be consistent with the society's definition of **ecotourism**. Another definition includes the idea of ecotourism being educational, raising environmental sensitivity, and altering behavior toward environmental soundness.[16]

Not all travel to natural areas is consistent with ecotourism or sustainable travel. An exhaust-spewing, petroleum-consuming, multithousand-ton cruise ship would not by any stretch of the imagination fit either definition even if it were plying the Amazon and the passengers were allowed ashore for shopping. By the same token, tiger hunting in Southeast Asia would not be considered ecotourism. Indeed, one of the major drawbacks of tourism to developing countries has been environmental degradation.[17] Consequently, the development of an ecotourism industry within any tropical country should involve careful planning and follow strict rules. Guidelines for developing ecotourism industries do exist, for example Butler and Waldbrook,[18] that are applicable to all parts of the planet.

It is technically arguable that true ecotourism is even possible. After all, even a troop of tourists who sleep in tents, replant trees, and live on nuts and berries they gathered themselves had to travel to their destination. Unless they walked and swam, their travel involved burning fossil fuels while riding on some form of conveyance that was constructed out of smelted ore. However, such semantic arguments ignore the idea that tourist money can have a beneficial effect on some impoverished parts of the world. For example, the village of Grande Riviere in Trinidad was a remote and impoverished community that for many years subsisted on cocoa farming. When the cocoa trees began to fail, it looked as if the village would go under. Tourism had never been part of the village's economy. However, in the early 1990s, for reasons that have not yet been clarified, leatherback turtles began nesting on the beaches in substantial numbers. This initially piqued local interest, but the word spread, and tourists began visiting from elsewhere in Trinidad and then from foreign countries. Interest in the turtles led to the building of inns for the tourists and employment for the villagers, which further led to conservation efforts and deliberate regeneration of the local environment. In general, it turned out to be positive for the local environment, the villagers, and the turtles.[19]

Sea turtle nesting has been responsible for conservation efforts along much of the Caribbean Coast of Central America, and tourism has grown as a result. In general, the efforts of people have contributed to the success of the turtles. It has not been an unqualified success, however. Tortuguero National Park in Costa Rica has drawn heavy tourist pressure as *gringos* have arrived in larger numbers than the turtles. The turtles continue to nest there successfully, but they tend to avoid the beaches on weekends when tourists are most numerous.[20]

Ecotourism certainly has the potential to generate income. According to Conservation International president Russell Mittermeier, it is, as of this writing, the fastest growing segment of the tourism industry, which is the fastest growing industry in the world.[21] One country that has been generally regarded as a pioneer in rainforest ecotourism is the Central American nation of Belize, located just south of the Yucatan in Mexico. Studies there have shown that ecotourism has led to local economic benefits and to generation of local support for conservation, as it did in Grand Riviere. On the negative side, it has not led to generation of financial support for the management of the protected areas, although the authors of the studies contend that it could with the implementation of user fees.[22] The lessons learned there could be applied to other countries interested in ecotourism.

Finally, perhaps the best model for ecotourism currently available would be found in the Central American nation of Costa Rica. It is reportedly earn-

Figure 12-2 Sea turtle depositing eggs on a beach. This is an annual event that attracts ecotourists to the Caribbean and elsewhere.

ing over $1 billion a year on ecotourism.[21] To have ecotourism, there has to be an ecosystem to tour, and Costa Rica apparently has made a commitment to keeping its rainforest intact. It pays landowners to refrain from deforesting their property. In this way, it becomes in the landowners best interest to not cut trees down. If a landowner harvests his trees, he makes money once; if he leaves them standing, he makes money as long as the government is willing to pay him. Reportedly this has not made much of an impact on deforestation rates in Costa Rica, but deforestation rates were already low when this program was enacted, as Costa Rica had implemented other conservation efforts.[23]

During the 1970s, before anyone had ever heard of ecotourism, Americans who wanted a rainforest experience used to travel to Suriname. It's about as far from the United States as is Costa Rica, and it is largely still forested. It does not have the stable political history that Costa Rica does, however, but it is starting to try to attract ecotourists. It also has another asset going for it with its pristine forests still in place: it can earn international money in carbon credits.[21] This is more appropriately discussed in the following chapter, but for the moment, suffice it to say that along with revenues from ecotourism, carbon credits can represent another financial incentive for countries with intact rainforests to preserve them.

In Closing

For those of who do not live in the rainforest, there are many reasons we can cite for its preservation. For the people who live there, however, especially those who live in poverty, the immediate needs for food, income, and even survival may make available rainforest resources too tempting to resist. Indeed, it is more in their self-interest to exploit these resources than to conserve them, even if it means rainforest destruction. To conserve the forests, it is necessary to gain the cooperation of these people, and that means making rainforest conservation profitable for them. Sustainable agriculture is one way, providing that those so engaged are able to make enough money from such work to make it worth their while to engage in it. Ecotourism is another, again if it is in the indigenous peoples immediate best interest. There are corporations and organizations that are working toward both ends, but they are fighting against long-established behavioral trends, and their task is formidable. Still, the goal is an admirable one and is well worth pursuing. One organization that is worth mentioning, but by no means the only one, is the Rainforest Alliance. Others include our familiar environmental organizations, but the Rainforest Alliance has specific programs in agriculture, ecotourism, and education. For anyone interested in learning more about the ways and means for rainforest conservation, the Alliance would be a good place to start. They're on the Web.

References

1. Diamond, J. 1994. Ecological collapses of past civilizations. *Proceedings of the American Philosophical Society* 138(3):363-370.
2. Diamond, J. 2000. *Ecological collapses of pre-industrial societies.* Stanford University: The Tanner Lectures on Human Values, May 22-24, 2000. http://www.tanner-lectures.utah.edu/lectures/Diamond_01.pdf.
3. Diamond, J. 2005. *Collapse: How societies choose to fail or succeed.* New York: Penguin Books.
4. Peterson, L. C., and G. H. Haug. 2005. Climate and the collapse of Maya civilization. *American Scientist* 93(4):322-327. http://www.americanscientist.org/template/AssetDetail/assetid/44510/page/1.
5. Tomich, T. P., M. van Noordwijk, S. A. Vosti and J. Witcover . 1998. Agricultural development with rain forest conservation: Methods for seeking best bet alternatives to slash-and-burn, with applications to Brazil and Indonesia. *Agricultural Economics* 19(1-2):159-174.
6. Clay, J. 2004. *World agriculture and the environment.* Washington, DC: Island Press.
7. Butler, R. 2005. Another look at global rain forest conservation: Are the rain forests still in need of saving? *Mangabay.com.* April 19, 2005. http://news.mongabay.com/2005/0419-rhett_butler.html.
8. Vanclay, J. K., and R. A. Preston. 1989. Sustainable timber harvesting in the rain forests of northern Queensland. *In:* Forest planning for people, *Proceedings of the 13th Biennial Conference of the Institute of Foresters of Australia*, Leura, NSW, September 18-22, 1989. Sydney, Australia: IFA, pp. 181-191.
9. Kammesheidt, L., P. Köhler, and A. Huth. 2001. Sustainable timber harvesting in Venezuela: A modeling approach. *Journal of Applied Ecology* 38(4):756-770.
10. Peters, C. M. 1994. Sustainable harvest of non-timber plant resources in tropical moist forest: An ecological primer. *People and Plants International.* http://www.peopleandplants.org/whoweare/Sustainability%20Primer.pdf.
11. Rain Forest Alliance. 1987-2227. http://www.rain forest-alliance.org/index.cfm.
12. Mann, C. C. 2002. The real dirt on rain forest fertility. *Science* 297(5583):920.
13. Nestle, M. 2007. Eating made simple. *Scientific American* September 2007, (297): 60-68.
14. Indonesian Mediawatch. 2007. *Losing Nemo.* Radio Singapore International, August 17, 2007. http://www.rsi.sg/english/indonesiamediawatch/view/2007 0820112652/1/.html.
15. The International Ecotourism Society. *Definitions and principles.* http://www.ecotourism.org/webmodules/webarticlesnet/templates/eco_template.aspx?articleid=95&zoneid=2.
16. Orams, M. B. 1995. Towards a more desirable form of ecotourism. *Tourism Management* 16(1):3-8.

17. Brohman, J. 1996. New directions in tourism for third world development. *Annals of Tourism Research* 23(1):48-70.
18. Butler, R. W., and L. A. Waldbrook. 2003. A new planning tool: The tourism opportunity spectrum. *Journal of Tourism Studies* 14(1):21-32.
19. Harrison, D. 2007. Cocoa, conservation and tourism Grande Riviere, Trinidad. *Annals of Tourism Research* 34(4):919-942.
20. Jacobson, S. K., and A. F. Lopez. 1994. Biological impacts of ecotourism: Tourists and nesting turtles in Tortuguero National Park, Costa Rica. *Wildlife Society Bulletin* 22(3):414-419.
21. Cairo, I. 2007. Suriname could regain leading position in ecotourism, says conservation official. *Caribbean Net News*, October 1, 2007. http://www.caribbeannetnews.com/news-3776—36-36—.html.
22. Lindberg, K., J. Enriquez, and K. Sproule. 1996. Ecotourism questioned: Case studies from Belize. *Annals of Tourism Research* 23(3)543-562.
23. Sánchez-Azofeifa, G., A. Pfaff, J. A. Robalino, and J. P. Boomhower. 2007. Costa Rica's payment for environmental services program: Intention, implementation, and impact. *Conservation Biology* 21(5):1165-1173.

The Role of Tropical Rainforests in Global Climate

13

A NUMBER OF YEARS AGO, I was invited to sit in on a student's presentation to the humanities faculty at the school were I was working on a paper she had written on tropical rainforests. The paper was well written and well presented, but it contained so many factual errors that I was appalled. Apparently the student had done the bulk of her research in the popular press and had neglected scientific journals. Two particularly egregious errors were that the rainforests are the "lungs" of the planet and that they're the world's principal defense against global warming. The two concepts are closely related, and the rainforests factor into them. But the idea that the rainforest is the major player in them is an enormous oversimplification. To consider the rainforest's role in global climate and climate change, we'll arbitrarily split the two concepts from one another and begin by exploring exactly what the lungs of the planet means and how the rainforests function in that capacity.

Lungs of the Planet?

Physiologically speaking, the lungs are the part of a terrestrial animal where carbon dioxide is exchanged for oxygen. Since an analogous organ for the planet is nonexistent, the concept is metaphorical, and it would apply to all of the planetary sites where carbon dioxide is exchanged for oxygen: the world's green plants.

Green plants exchange carbon dioxide as a result of **photosynthesis**, the biochemical process in which atmospheric carbon dioxide reacts with hydrogen from water absorbed from the ground to synthesize glucose, an elementary sugar from which most biological molecules are ultimately derived. The process is summarized by the chemical equation

$$6CO_2 + 6H_2O \Rightarrow C_6H_{12}O_6 + 6O_2 \uparrow$$

where CO_2 represents carbon dioxide, H_2O represents water, $C_6H_{12}O_6$ represents glucose, and $O_2 \uparrow$ represents the oxygen that is evolved back into the air as a waste product of the process. In order for photosynthesis to occur, the green pigment **chlorophyll** must be present as must light. Additionally, the ambient temperature must be greater than 0°C. Frozen water will not work in photosynthesis. Consequently, as winter approaches, photosynthesis slows down and ultimately stops until warmer temperatures return in spring. In terrestrial ecosystems, green plants carry on the overwhelming majority of all photosynthesis; a minor amount is accomplished by lichens and algae. In aquatic and marine ecosystems, algae are the dominant photosynthesizers, although photosynthesis is also carried on by some bacteria.

Two characteristics of tropical forests would give the impression that they are the major sites of photosynthesis on the planet. First, the ambient temperature never drops below freezing. Therefore, water is always available and photosynthesis can occur during the daylight hours every day. Second, the sheer bulk of green plants on an area basis is probably greater in the rainforests than in any other terrestrial ecosystem. Consequently, logic would indicate that the rainforests must be the planet's greatest site for photosynthesis.

A graph showing the annual levels of atmospheric carbon dioxide (Figure 13-1) for the last 50 years shows that the absolute level has been rising, which is consistent with the global warming hypothesis and will be discussed in the next section. It also shows that there is considerable fluctuation every

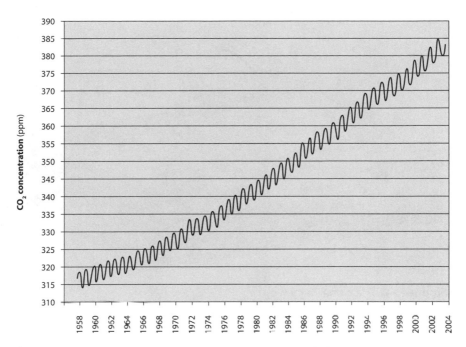

Figure 13-1 Chart of annual atmospheric carbon dioxide flux since 1960. Data courtesy of Scripps CO2 Program.

year, with the high point occurring during the northern hemispheric winter and the low point during the summer. Essentially, this means that when photosynthesis ceases as the northern hemisphere cools, less and less carbon dioxide is removed from the atmosphere. However, the forces that produce carbon dioxide continue to function. Moreover, rainforests occupy only 7% of the planet's surface at best. Given that the greatest concentration of land on the planet lies in the North Temperate Zone, it would make sense that this is where a significant amount of photosynthesis would occur. Consequently, much of the carbon dioxide that is removed from the atmosphere is done so in the Northern Hemisphere.

There are two other aspects to atmospheric carbon balance: **respiration** and **decomposition**. While plants remove carbon dioxide from the air when they photosynthesize, they consume oxygen and contribute carbon dioxide to the air when they carry on respiration, like animals (Figure 13-2). Moreover, when bacteria break down or decompose organic matter of all kinds, they often consume oxygen and liberate carbon dioxide as well. In overall effect, the reactions of respiration and decomposition are the reverse of photosynthesis, and while more photosynthesis per unit area may occur in tropical rainforests, more respiration does as well, and it occurs all year.

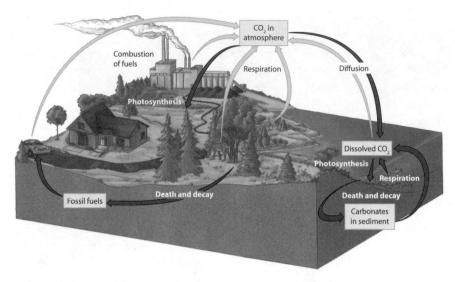

Figure 13-2 The global carbon cycle.

This is not to suggest that tropical forests are unimportant in global atmospheric carbon dioxide fluctuation, but the function of other ecosystems is important as well. Moreover, much carbon dioxide is absorbed and dissolved by the world's oceans, a phenomenon that is generating concern over changes in the oceans' chemistry. It will be discussed in another book in this series.

When a forest is disturbed, it will often eventually return to its undisturbed state by the process of succession, if given the opportunity and if disturbance is not too severe. Depending upon the extent of disturbance and type of forest, the initial or pioneer plants in the successional process are usually soft stemmed, but they are eventually replaced by woody plants such as bushes and shrubs, which, in turn, are eventually replaced by trees. As trees grow, much of the carbon dioxide that is fixed during photosynthesis is sequestered in the tree in woody tissue, where it may be held for hundreds of years. Once the tree dies, it will decompose, of course, and the carbon dioxide trapped in its tissues will be returned to the atmosphere, but very slowly. In contrast, soft stemmed plants are ephemeral and are easily decomposed after they die; most of the carbon dioxide they absorb during a growing season will be returned to the atmosphere within a few years after the plant dies, although some may be sequestered in the soil as organic material or humus. Leaves of trees act in the same manner. In essence, then, the more woody tissue that is produced, the more carbon dioxide is removed from the atmosphere.

In most tropical rainforests, little, if any, organic material tends to accumulate in the soil. With peat forests being an exception, dead herbaceous plants and fallen leaves decompose very quickly. The year-round mild to warm temperatures in rainforests encourage the activity of decomposing bacteria. In contrast to temperate forests, tropical forests have very little leaf litter on their floors (Figure 13-3). Furthermore, as a forest matures, respiration tends to catch up with photosynthesis. At one time it was believed that as a forest matured, it eventually reached a point at which succession ceased and the forest remained unchanged from that point on unless it was disturbed by some outside force. This was referred to as a **climax** forest.[1] In the climax forest, there was no net growth, and all photosynthesis was offset by decomposition and respiration. Many contemporary ecologists do not like the concept of a climax forest because it implies stagnation, which is clearly not the case. Although an old growth forest may appear to be unchanging to the untrained observer, disturbance is very much a part of forest ecology, and dynamic forces are very much at work. Trees do die and fall over, for example, or the weather may be unusually severe in a given year, and the organisms must make appropriate adjustments to the changing conditions. Still, the amount of change that occurs in a mature forest is small in contrast to a younger forest where trees are growing rapidly. As a result, middle-aged forests sequester more carbon than do fully mature forests.[2] Disturbance of old growth forests does release large quantities of carbon dioxide into the

Figure 13-3 Floor of tropical rainforests showing leaf litter.

atmosphere over a short period of time. In addition, changes in rainfall patterns brought on by forest clear-cutting exacerbate the problem by promoting fire, as has happened in Brazil.[3] Therefore, protection and preservation of standing forests is important in terms of not adding to the carbon dioxide load of the atmosphere. However, if there is any chance of actually reducing that atmospheric load, it would seem that massive reforestation would be an important part of the process.

In summary, the long-term removal of carbon dioxide from the atmosphere by photosynthesis is a complex process. All green plants are capable of absorbing carbon dioxide from the air. Keeping it out of the air is another matter. Tropical rainforests certainly contribute to carbon dioxide sequestration, but they are by no means the only ecosystems involved in the process. Consequently, while metaphors such as "lungs of the planet" may be attractive and easy to understand, they are oversimplifications and of questionable accuracy. Worse, they can invite ridicule, for example see Scott,[4] and in the long run may add more to confusion than enlightenment.

Global Warming

Throughout history, science and popular culture have often come into conflict. In the 1600s, Copernicus and Galileo got things going by demonstrating that the Earth was not the center of the universe. Two centuries later, Charles Darwin and Alfred Russell Wallace really fanned the flames with their theory of evolution, for which Darwin gets the lion's share of the credit. That particular argument rages even today, although ironically in the United States, not in England where it was originally published. In both of those cases, there were other scholars who contributed to the advancement of knowledge beyond those credited, and that is probably the case with global warming as well, the current bone of contention. One name that is certainly associated with the dissemination of information on global warming is that of former US Vice President Albert Gore. Gore's motion picture on global warming, *An Inconvenient Truth*, having earned him a Nobel Prize and Academy Award, will probably also earn him a place in history for having made information on the subject accessible to and understandable for so many people. But despite his efforts and those of the scientists who have been working on the subject, there is still much resistance to accepting global warming as a planetary threat. Much of it comes from radio talk show hosts and politicians. And while scientists may not be totally unanimous in their opinions about global warming, most agree that the available evidence supports the conclusion that the planet is getting warmer (Figure 13-4), and that rising carbon dioxide levels in the atmosphere are contributing to it.

The phenomenon behind global warming has been called the **greenhouse effect** (Figure 13-5). Essentially, this is explained as atmospheric gasses such as carbon dioxide function similarly to a pane of glass in a greenhouse, allowing visible sunlight to penetrate the atmosphere and strike the surface of the earth. Some of that light is absorbed, converted to infrared light—heat—and reradiated back toward space. However, the greenhouse gasses, again like a pane of glass, are opaque to infrared light; they reflect it back into the atmosphere, warming the planet. The moon, which receives the same concentration of sunlight as the Earth, lacks an atmosphere and so it does not warm.

For just about all of the planet's history, greenhouse gasses were balanced by natural phenomena. Studies of ice cores retrieved from the Greenland ice cap have shown that colder periods in Earth's history coincided with lower atmospheric CO_2 levels, but other phenomena such as atmospheric debris, as from volcanic eruptions, and fluctuations in sunlight intensity have been factors as well. Most recently, however, the warming of the planet is most probably the result of the loading of the atmosphere with greenhouse gasses.

Examination of Figure 13-2 shows that among the sources of carbon dioxide now contributing to the atmospheric load is the burning of fossil fuels. Prior to the industrial revolution this probably amounted to a trivial

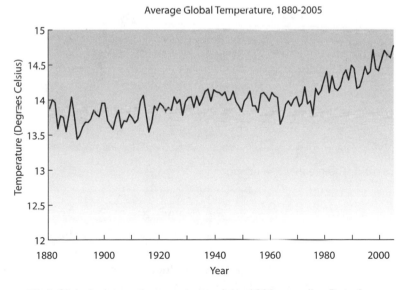

Figure 13-4 Global average temperatures since 1960 or earlier. Data from Goddard Institute for Space Studies, NASA Goddard Space Flight Center, *Earth Sciences Dicectorate*, "Global Temperature Anomalies in .01 C, March 2006."

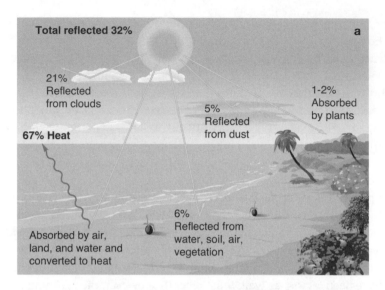

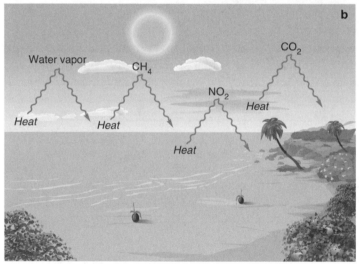

Figure 13-5 The greenhouse effect.

contribution. Since then, tons of carbon have been added to the atmosphere by the combustion of petroleum, coal, and natural gas. The reason for this is simple enough: carbon dioxide is a by-product of combustion. Air pollution controls may be capable of taking such combustion products as sulfur dioxide out of fossil fuel combustion emissions, but there is simply no way to remove carbon dioxide.

The use of biogenerated fuels, particularly ethanol, as an alternative to fossil fuels to reduce carbon dioxide emissions has become an attractive argument. In actuality, it has been going on in Brazil since the 1970s, and according to two University of Sao Paulo scientists, Jose R. Moreira and Jose Goldemberg, until at least 1997, it has been largely successful.[5] In Brazil, ethanol is generated from sugarcane; about 65% of the total crop reportedly is used that way. According to the authors, it has largely been a successful program, not only relieving Brazil of the costs of importing fossil fuels, but also in terms of social improvements to the lives of the sugarcane workers and reductions in air pollution, including carbon dioxide emission. Moreover, as sugarcane production has expanded, improvements in agricultural techniques and plant genetics have actually increased yield. And while the program was at first very expensive with ethanol from cane initially costing around $100 per barrel, greater production and technical advances have reduced costs.[6] With the recent increases in petroleum prices worldwide, ethanol may be becoming competitive with conventional fuels in Brazil.[5,6]

In terms of the impact of ethanol production on the rainforest, it's too early to draw much of a conclusion, and predictions are always speculative. In Brazil, expanded sugarcane cultivation has largely been at the expense of coffee production.[5] However, as petroleum prices continue to increase, ethanol production will become more and more attractive, and tropical countries may shift more agriculture toward cane production. Conceivably, the rush to produce alcohol from sugarcane could accelerate rainforest destruction.

One note on this topic regarding ethanol production in the Northern Hemisphere: the Brazilian experience to date may not be applicable. First, Brazil is not one of the world's major sugarcane producers, and its conversion of cane from sugar to alcohol production won't have much affect on world prices. The United States appears to be basing much of its ethanol production on corn, and as more U.S. corn is shifted from consumption to ethanol production, global corn prices will be affected, as the United States is the major producer. Second, the production of ethanol from sugar cane appears to be a largely efficient process in that more available energy is produced than is consumed by production. According to at least one study, although others disagree, ethanol production from corn actually consumes more energy than it produces.[7]

It is important to remember that carbon dioxide is only one of a number of gases involved in global warming, although it is the most abundant. There are several other greenhouse gasses as well, including methane (CH_4), the dominant component of natural gas and a far more potent greenhouse gas than carbon dioxide. Methane is emitted into the atmosphere naturally by termite nests, **anaerobic** bacterial decay, wetlands like swamps and marshes, forest and grass fires, and even standing forests.[8] Global warming deniers are quick to point this out. However, there are numerous **anthropogenic** sources of methane as well. They include rice paddies, cattle and sheep, fossil fuel burning, gas flaring, coal mining, and landfills, and they account for approximately 60% of the atmospheric methane sources.[9] Natural sources have existed as long as life has and have been balanced by natural methane removal from the atmosphere. Anthropogenic sources have added to the atmospheric load, and there is now more methane in the atmosphere than there has been at any time in the past.[9]

Rarely mentioned is the role that human population size plays in the production of greenhouse gases . While gross emissions have grown since the industrial revolution, emissions per person have increased as well. Moreover, it is human population growth that has contributed to more rice paddies, more methane-emitting domestic livestock, and more fossil fuel consumption, while simultaneously contributing to forest destruction. Consequently, even if emissions per person are reduced, the immense size of the global human population will probably guarantee that total greenhouse emissions will remain high.

Most likely, the complete understanding of global warming lies in our future, and by the time it's accomplished, it may well be too late to avoid a considerable amount of damage to the planet. Human life will be affected, although there continue to be those who remain in denial. In fact, there is an argument that the increasing carbon dioxide levels in the atmosphere are really beneficial, because they will promote more plant growth. So far, scientific research has not supported this.[10]

Carbon Credits

One of the proposed solutions to global warming is the sale of "carbon credits." As I understand this system, there are calculated limits to how much carbon dioxide a country can emit. Countries such as Suriname that are emitting less than their quota can sell credits to parties, countries, companies, even individuals, who are exceeding theirs. Thus, the forested countries have a financial incentive to keep their emissions down, presumably by preserving their forests and not developing industrially. Moreover, it gives poor countries, presumably tropical ones, a financial incentive to reforest while industrial parties are able to continue doing business as usual, presumably with the understanding that they will reduce their emissions as technology improves and it becomes financially more feasible. Since I am not an economist, I cannot comment on the efficacy of this practice. As a citizen of the planet, I hope it works. From my reading on the subject, however, I have found that there are those in the media and elsewhere that express serious reservations.

In Closing

Although popular perception of global warming has it that tropical defor-estation is a major factor,[11] it appears to be for the wrong reason. Tropical forests do not sequester carbon dioxide as most people think. Rather, it's their destruction that releases massive amounts of stored carbon into the air. Most people do appear to be aware that at least part of the problem with global warming is the increasing amount of atmospheric carbon dioxide. Many would like to do something about it, but they don't know what.

This is not to suggest that there is universal agreement on the subject. Many people, including some that are influential, continue to deny that global warming is a problem. Some claim that it is simply a natural climatic fluctu-ation; others call it an outright hoax. Indeed, in what can only be described as utter absurdity, some people claim that global warming is a fabrication of politicians or that there is serious disagreement among scientists. In reality, much of the opposition to global warming, in the United States anyway, comes from politicians or those who are politically aligned, while scientists generally, though not totally, support the idea. It is beyond dispute that the planet is getting warmer; what disagreement exists is over how much is the fault of humans and how bad the eventual outcome will be.

It's difficult to predict what the consequences of global warming will be. Coastal flooding caused by increases in ocean levels due to the polar ice caps melting, more and longer dry seasons promoting more brush and forest fires, and more severe storms have been predicted. To some extent, we are seeing some of that now. Forest fires in the western United States have been grow-ing more and more intense. Most scientists appear to believe that at worst, civilization will be brutally disrupted, but the life-support systems of the planet will be maintained. Still, not all agree. A. V. Karnaukhov of the Russian Academy of Sciences, for example, has written that human activities will so disrupt the climate that natural forces that remove carbon dioxide from the atmosphere will be overwhelmed, and temperatures will rise to points greater than those compatible with life as we know it.[12] At this point it is impossi-ble to know who is correct.

To those who wish to pursue this subject further, I would suggest be-ginning with the Intergovernmental Panel on Climate Change (IPCC), who shared the recent Nobel Peace Prize with Al Gore. They are on the web at http://www.ipcc.ch/. From there one can search a variety of sources, but it is important to remain objective, because not all sites are equally reliable.

References

1. Snyder, M. 2005–2007. Wood whys: What is a climax forest. *Northern Woodlands* http://www.northernwoodlands.com/wood_whys.php.
2. Pregitzer, K. D., and E. S. Euskirchen. 2004. Carbon cycling and storage in world forests: Biome patterns related to forest age. *Global Change Biology* 10(12): 2052–2077.
3. Wallace, S. 2007. Farming the Amazon. *National Geographic* 211(1):40–71.
4. Scott, P. 1999. *Tropical rain forest: A political ecology of hegemonic myth making.* London, Institute of Economic Affairs. http://www.iea.org.uk/files/upld-publiction2pdf?.pdf.
5. Moreira, J. R., and J. Goldemberg. 1999. The alcohol program. *Energy Policy* 27: 229–245. http://www.unb.br/fav/renova/reno/Artigos/alcool_Br.pdf.
6. Goldemberg, J., S. T. Coelho, P. M Nastari, and O. Lucon. 2004. Ethanol learning curve—The Brazilian experience. *Biomass and Bioenergy* 26(3):301–304.
7. Pimentel, D., and T. W. Patzek. 2005. Ethanol production using corn, switch-grass, and wood; Biodiesel production using soybean and sunflower. *Natural Resources Research* 14(1):65–76.
8. Keepler, F., and T. Röckmann. 2007. Methane, plants and climate change. *Scientific American* 296(2):52–57.
9. Anon. 2006. Methane as a greenhouse gas. *CCSP Research Highlight 1 (January 2006).* http://www.climatescience.gov/infosheets/highlight1/default.htm.
10. Feeley, K. J., S. J. Wright, M. N. Nur Supardi, A. R. Kassim, and S. J. Davies. 2007. Decelerating growth in tropical forest trees. *Ecology Letters* 10(6):461–469.
11. Bord, R. J., A. Fisher, and R. E. O'Connor. 1998. Public perceptions of global warming: United States and international perspectives. *Climate Research* 11:75–84. http://www.int-res.com/articles/cr/11/c011p075.pdf.
12. Karnaukhov, A. V. 2001. Role of the biosphere in the formation of the Earth's climate: The greenhouse catastrophe. *Biophysics* 46(6):1138–1149. http://www.poteplenie.ru/doc/biophis-eng.pdf.

Tropical Dry Forests

14

THE ULTIMATE SOURCE OF WATER for the tropical rainforests is the world's oceans. Although half of the rainfall that occurs there is the result of transpiration, without water vapor from the ocean, there wouldn't be enough rain to support the wet tropical forests. Indeed, as tropical forests become isolated from the ocean, either by distance or by physical barriers such as mountains that the water vapor can't cross, rainfall is reduced and the nature of the forests changes. The amount of rainfall, of course, is only one criterion in shaping an ecosystem. Temperature is another, and so is the seasonal distribution of the rainfall. Extreme seasonal rains triggered by monsoon winds can be massive and generally support a tropical forest that differs from the typical tropical rainforest. But it is still considered to be wet, or at least seasonally wet.

Not all tropical ecosystems and biomes are wet, of course. Deserts are characteristically dry biomes that are dominated by plants that are able to store water and/or are anatomically adapted to prevent water loss. Such plants may also have deep roots that are able to penetrate far below the ground surface where water may be found, and they are biochemically adapted to carrying on photosynthesis in dry or **xeric** conditions. Cacti are the familiar plants of the American hot deserts, of course, but numerous grasses and shrubs are found in deserts as well. Tropical steppes and savannas are dry biomes that are dominated by grasses, although savannas are grasslands characterized by trees growing singly or in groups punctuating the landscape. These arid biomes will be considered in another book in this series; they are mentioned here only for contrast to one additional dry tropical ecosystem: the tropical dry forest. Also known as the tropical seasonal forest or tropical deciduous forest, this biome is dominated by **deciduous** trees that drop their leaves during the dry season. There is at least one dry season each year, usually from two to three months in duration, but there may be a second, minor dry period as well. The greater the distance between the biome and the ocean, the longer the dry season will be[1] (Figure 14-1).

Dry tropical forests are distributed around the globe. They occupy much of Africa south of the Sahara and north of the Tropic of Capricorn, outside of the African rainforest, of course, as well as India and much of mainland Southeast Asia inland from the coast. It is also found in northern Australia.

Figure 14-1 Dry tropical forest. Note the abundant leaf litter on the forest floor, unlike that in rainforests.

In the Western Hemisphere, it can be found in Mexico, down the west coast of Central America into Costa Rica, and in South America along the northern coasts of Columbia and Venezuela and in Brazil southeast of the rainforest (Figure 14-2).[2] Some exist as **rain shadow** islands within the Andes.

In general, dry tropical forests are more tolerant of disturbance than are wet forests. If they are disturbed, they tend to recover more quickly. This assumes, of course, that disturbance is not ongoing. People use dry forests as sources of firewood and charcoal as well as for planting and grazing of livestock. Where population pressure is heavy, the dry forest is as threatened as is the rainforest. Oddly enough, however, fire is not a normal threat to the dry forest. When it does occur, it is more likely caused by some human activity than a natural one.[1]

Soils

Because of the alternating wet and dry seasons. Soils dynamics of dry tropical forests are different than those of the wet forests. On the whole, soils of dry forests are more fertile as nitrogen and potassium and, to a lesser extent, phosphorus tend to be stored there. Deciduous plants withdraw phosphorus from the leaves before they're shed at the onset of the dry season. But leaves emerge prior to the onset of the rainy season, when nutrients are liberated from soil particles, and there is less leaching of nutrients from the soils.[1]

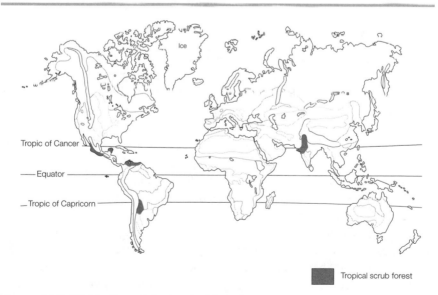

Ice

Tropic of Cancer

— Equator —

— Tropic of Capricorn —

Tropical scrub forest

Figure 14-2 Distribution of dry tropical forest.

When nutrient availability and plant uptake coincide, the nutrients tend not to be lost.[3] In all, not only is the fertility of the soils greater in the dry forest than in the rainforest, but the dry season appears to make the dry forest better suited for agriculture than is the rainforest.[1,4]

Plants

Not surprisingly, the dominant plants in the dry forest are the trees. In general they are smaller in size and diversity than are those of the rainforest, and they tend to be deciduous, losing their leaves during the dry season. In fact, some species of trees common to both the rainforest and dry forest are evergreen in the one and deciduous in the other.[5] Stratification does occur, though again not to the extent that it does in the rainforest. There is a canopy, usually between 10 and 30 meters high, with emergents. Lianas are less abundant, and the understory, which consists of a subcanopy layer and low shrubs, is dense and tangled. There is also growth on the ground.[5,6]

Trees tend to have adaptations that help them resist desiccation. Their bark is thicker and more ridged, and they may have water storage capability in their tissues, particularly their roots. They also send their roots deeper into the soil. During the dry season, plants may not necessarily go dormant; they may have chlorophyll in or under their bark, which allows the stem or trunk to carry on photosynthesis even in the absence of leaves. An example would be the *Bursera simarouba* of Central America, commonly called naked Indian or sunburned tourist because of its peeling bark. Other adaptations include folding their leaves at night, much as a flower closes, to conserve water.[6,7]

The most common family of plants in the African and American tropical dry forests is the legumes.[8] Among them are numerous species of *Acacia*, some of which house symbiotic ants as do some rainforest representatives of the genus. A unique dry forest exists in Belize, Central America, that is dominated by Caribbean pine. Other trees are present, including some oaks, palms and palmettos, and cecropias, a tree found in rainforests. Most dry forests are fairly diverse, though less so than the rainforests, but this one is quite simple in contrast, and fire is sometimes a factor in this unique ecosystem.[9,10]

As mentioned earlier, in Africa the dry forest covers much of the continent. Africa is also overrun by poverty, disease, overpopulation, and civil unrest, and the dry forest has taken something of a beating. However, the Center for International Forestry Research points out that there are resources in the African dry forest that if managed properly could very well contribute to better living standards for the people there.[11] An example would be the shea tree, *Vitellaria paradoxa*, which is being managed for conservation, food, and income in Uganda. The income generated from the sale of shea butter, a fatty material extracted from the tree, is used to improve the lives of the Ugandans who are growing it. Shea butter is used in the United States as a cosmetic agent, similar to cocoa butter.[12]

As is the case with the tropical rainforest, the tropical dry forest is home to a variety of epiphytes (Figure 14-3). They include orchids and bromeliads, as well as cacti and aroids.[4] The last mentioned are from the plant family Araceae, which includes the tropical ornamental *Philodendron* plus some familiar North American plants such as the Jack-in-the-pulpit and skunk cabbage. Some of these plants survive the dry season by storing water in specialized tissues. Bromeliads store water around the base of their leaves and often act as aquaria, as it were, for small, arboreal animals.[4]

Animals

According to Kricher, animal diversity is less in the dry tropical ecosystems than in the rainforests, but there is still a considerable variety present. Moreover, one can often find rainforest animals in the dry forests and savannas, particularly where the ecosystems are adjacent to each other, although the rainforest animals do not wander far from their preferred habitat. In Latin America, anteaters and giant armadillos enter dry ecosystems to search for ants.[9] Depending upon where in Latin America a particular dry forest happens to be, one may find different varieties of monkeys, deer, javelinas, coatis, and small cats.[4] In Africa, animals of the dry forest include the exotic, charismatic megafauna that one typically thinks of when African animals are mentioned: giraffes, lions, and large herds of grazing animals.[11]

The tropical dry forest also houses numerous birds, including some that are endemic to the ecosystem, some that inhabit it and other ecosystems, and some that are seasonal visitors. D. L. Pearson found that more than three fifths of the birds found in a dry forest in Eastern Peru were also found in Suriname,[6] much of which is rainforest, and Herzog and Kessler found that almost

Figure 14-3 Dry forest epiphytes.

80% of bird species in isolated dry forests in Bolivia were found in other ecosystems.[13] Trogons and motmots, species I've seen in the rainforests of Costa Rica, also occur in some dry forests there and elsewhere in Latin America. Yellow-bellied sapsuckers and house wrens, which spend their summers in North America, migrate to South and Central America for the winter and occur in dry forests.

Reptiles of the dry forests include the usual varieties: snakes and lizards. Again, one finds species that are also found in nearby rainforests. In tropical America, this includes boa constrictors, geckos, and iguanas. Even amphibians can be present. Many of these probably wander in from adjoining rainforests during the wet months, but many may survive in wet areas along the banks of streams, even seasonal ones, and others may take to water reserves in bromeliads. A particularly interesting amphibian that can be found in dry forests is the marine toad *Bufo marinus*. This highly adaptable animal has been introduced into tropical and semitropical areas around the world in an attempt to control insect pests, though not very successfully. Instead, it has become established as an invasive species, often outcompeting native amphibians and causing injury to people and pets because of a toxin it secretes. It even robs cat and dog food that people have put out for their pets.[14]

As with most other ecosystems, the most abundant and diverse animals in the tropical dry forest are the arthropods. A large variety of insects and other small arthropods exist in the soil and on plants, particularly within the canopy. Ants are particularly abundant and diverse, some of them living mutualistically with acacias, as is the case in the rainforests. In addition, there are the more fascinating, if scary, members of the phylum: scorpions and tarantulas. In particular, the Mexican true red leg is a large tarantula that some people actually keep as pets. It is not particularly dangerous, but it is impressive. According to Costa Rican guide and naturalist Gilberth Calvo, the scorpions of the arid Central American environments present more of a danger to humans than do those of the wet forests. Both animals are nocturnal predators that hunt insects and other invertebrates, although the tarantula may take small vertebrates if the opportunity presents itself.

Conservation Issues

Most North Americans, at least north of Mexico, are unaware of the existence of dry tropical forests. Many are aware of the rainforests, of course, and the threats that are facing them, but most have never heard of dry tropical forests let alone that they are undergoing deforestation more rapidly than are the rainforests. Tropical dry forests around the world are severely threatened. Less than 1% of the original Pacific Coast dry forests of Central America remain, for example.[6] Indeed, they have been described as the most highly threatened tropical forest habitats on Earth, largely because they are easily cultivated.[15] More recently, particularly in Africa, the dry tropical forests are coming under more and more pressure as people comb them for fuel wood. Moreover, there and in India, overgrazing adds additional pressure, not only to dry forests but to arid woodlands and grasslands in general. Consequently, as population pressures continue in some parts of the world, and economic pressures continue in others, more destruction of these ecosystems are likely to occur. Scientists are aware of the problems, and technology exists to solve them.[16] The question now is whether or not the will to do so exists.

References

1. Murphy, P. G., and A. E. Lugo. 1986. Ecology of tropical dry forest. *Annual Review of Ecology and Systematics* 17:67–88.

2. Marietta College Department of Biology and Environmental Science. Undated. *Biomes of the world: The tropical seasonal forest.* http://www.marietta.edu/~biol/biomes/tropdry.htm.

3. Lodge, D. J., W. H. McDowell, and C. P. McSwiney. 1994. The importance of nutrient pulses in tropical forests. *Tree* 9(10): 384–387. http://www.fpl.fs.fed.us/documnts/pdf1994/lodge94a.pdf.

4. Neill, D. 2000, *Observations on the conservation status of tropical dry forest in the Zapotillo area, Loja Province, Ecuador.* Missouri Botanical Garden. http://www.mobot.org/MOBOT/research/ecuador/zapotillo/report.shtml.

5. Slater Museum of Natural History. 2007. *Tropical dry forest.* University of Puget Sound. http://www.ups.edu/x6105.xml.

6. Pearson, D. L. 1971. Vertical stratification of birds in a tropical dry forest. *Condor* 73:40–55. http://elibrary.unm.edu/sora/Condor/files/issues/v073n01/p0046-p0055.pdf

7. Ceiba Foundation for Tropical Conservation. 2007. *Lalo Loor dry forest.* http://www.ceiba.org/loor.htm.

8. Olson, M. E. 1999. *Images of dry tropical habitat: the legume family (Fabaceae).* http://www.mobot.org/gradstudents/olson/legume.html.

9. Kricher, J. 1997. *A neotropical companion.* Princeton, NJ: Princeton University Press.

10. Andraka, S. C. Locklin, and J. Schipper. 2001. *Belizean pine forests (NT0302).* World Wildlife Fund. http://www.worldwildlife.org/wildworld/profiles/terrestrial/nt/nt0302_full.html.

11. Center for International Forestry Research. 2007. *Africa's dry forests.* http://www.cifor.cgiar.org/Publications/Corporate/NewsOnline/NewsOnline39/dry_forests.htm.

12. Anon. Undated. *The Shea project.* http://www.thesheaproject.org/index.html.

13. Herzog, S. K., and M. Kessler. 2002. Biogeography and composition of dry forest bird communities in Bolivia. *Journal für Ornithologie* 143(2):171–204.

14. Higris, R. 2001. *Bufo marinus*. University of Michigan Museum of Zoology Animal Diversity Web. http://animaldiversity.ummz.umich.edu/site/accounts/information/Bufo_marinus.html.

15. Janzen, D. H. 1988. Management of habitat fragments in a tropical dry forest: Growth. *Annals of the Missouri Botanical Garden* 75(1):105–116.

16. Bellefontaine R., A. Gaston, and Y. Petrucci. 2000. *Management of natural forests of dry tropical zones. FAO Conservation Guide* 32. http://www.fao.org/docrep/005/W4442E/w4442e00.htm#Contents.

What Comes Next?

15

ROUGHLY 65 MILLION YEARS AGO, an asteroid perhaps 10 kilometers across struck the Earth near what is now the Yucatan peninsula of Mexico. The impact kicked tons of debris into the air, blocking out sunlight around the world and causing a drop in temperature that lasted for months if not years. That event is thought to have brought about the extinction of the dinosaurs and 70% of all other species on Earth. Life on the planet was forever changed. Had that event not happened, we probably would not be here today.

Five other mass extinctions are thought to have occurred during the history of life. Some scientists argue that a sixth is going on right now, and we are causing it. Given that the tropical forests are the greatest reservoirs of biodiversity on the planet at the moment, if we succeed in destroying them, then by definition we will have brought about a mass extinction. The questions then become: what impact will it have on us, and what can we do to prevent it?

Humans and Extinction

There is nothing new or unusual about extinction. Easily over 90% of the species of organisms that have ever lived are now extinct, and all now living will eventually become extinct, ourselves included. The difference between past extinctions and those that occurred in the last 10,000 years or so is that humans, it has been argued, have caused many of the latter. Indeed, it appears that extinctions of animals in any given area have coincided with human arrival. On many islands, for example, birds and reptiles disappeared when humans settled. Many large animals disappeared from North America when humans first arrived, although human arrival coincided with the end of the last ice age. Climate change cannot be ruled out in the extinction of some of those animals. Humans arrived in North America when the great glaciers of the time were starting to retreat and the climate was warming. Still, humans have pretty well been implicated in the extinction of the North American ground sloths,[1] for example, and more recent extinctions, such as the passenger pigeon, have been clearly human caused.

Extinction can come about in a number of different ways. Overhunting is probably the first to come to mind, and in the case of the ground sloth, it is the probable reason. However, other ways include competition for resources, such as food and living space, habitat destruction, and introductions of competitors, predators, and parasites. As described earlier, as contemporary humans have moved deeper and deeper into tropical forests, they have crowded out other organisms, including indigenous people already living there; cleared or otherwise changed the forests for their own use; and introduced their own plants and animals. Some native species have been able to withstand the onslaught; others have not. Many of the species lost in rainforest destruction were undescribed; consequently, any potential use they may have had for us has been lost as well, and changes that have been forced on indigenous peoples have caused the loss of knowledge on the use of native species, medically and otherwise.[2]

For several years there has been disagreement over the best way to manage the planet, with those advocating exploitation as always maintaining the upper hand. However, more and more people are coming to realize that business as usual will ultimately lead to the inability of human society to maintain itself, and many have begun efforts to effect change. In terms of protecting the tropical forests, there are two principal ways: conservation and reforestation.

Conservation

Throughout human history, nature was seen in two ways: as something to be exploited and used to our advantage, and as an adversary that was to be conquered. Natural resources were plundered mercilessly until they became scarce, and then they were sought somewhere else; rivers were dammed, channelized, and diverted, all for human benefit. Unfortunately, trying to manipulate nature is something akin to squeezing a water balloon; when it's constricted in one place, it bulges somewhere else. Thus, exploited resources often meant scarred and eroded land, destroyed forests, pollution, and extirpated populations of plants and animals. In some cases, the remnants of the resources have become harder and harder to exploit. Dammed, diverted, and channelized rivers often caused downstream erosion of riverbeds, silted reservoirs, and ironically, greater flooding.

Unfortunately, not everyone has been able to learn from bad experiences; there are those who still insist that we continue dealing with nature as we have in the past and that somehow everything will take care of itself. Others believe that an environmental holocaust is about to strike unless we change our behavior radically. To both groups, conservation seems to mean hands off: remaining resources, especially **biotic resources**, should be left alone. Those who favor exploitation are generally antagonistic to conservation and its proponents. In practice, however, conservation of natural resources does not necessarily mean not using them. That's **preservation**. Conservation means the careful use of resources. In the case of **abiotic resources** such as minerals and ores, that would mean using them in such a way to minimize or, preferably, eliminate waste. It would encourage the recycling of those resources wherever possible once their useful life was over. In the case of biotic resources, conservation would be more like sustainable agriculture or forestry. Use should be no greater than natural replacement.

Conservation of ecosystems such as tropical rain and dry forests is often accomplished by setting aside tracts of land to be left undisturbed, essentially parks or preserves. Use of such lands for recreation and environmental research is usually permitted, but any kind of destructive exploitation such as logging or hunting is not. In theory, such preserves will maintain reservoirs of species and ecosystems. In practice, it's more complicated.

Whenever an ecosystem is disrupted in such a way that isolated parts of it remain undisturbed, **fragmentation** is said to have occurred. The fragments are essentially islands of the original ecosystem that are surrounded by destruction, which may be a barrier to the movement of animals and plants between the fragment and continuous forest. Plant movement would of course be the carriage of seeds or pollen by animal symbionts. Usually the transition from disturbed to undisturbed forest is abrupt, rather than the gradual transition or **ecotone** one would find in a transition zone between two natural but dissimilar ecosystems. The abrupt edge (Figure 15-1) would expose mature trees to wind, driving rain, and temperature stresses they normally do not endure within the sheltered forest, and the so-called edge effect can be harmful. Consequently, trees experience structural damage and dehydration. With more sunlight available, vines, lianas, and successional plants grow along the edge making it look dense and impenetrable. Size and shape of the fragment will also make a difference in terms of how it will be affected. Long, narrow fragments have relatively more edge than do rounder ones of equal area, and smaller fragments undergo greater change and destruction than do larger ones.[3,4] In other words, one plot of 500 acres would conserve more biodiversity than 500 plots of one acre each.

Part of the problem with forest remnants is the dependence many plants have on specific animal pollinators and seed dispersers. Euglossine bees, for example, as well as a number of small animals and understory birds will not cross cleared areas; they will function only in continuous forest.[4] If there is no hive of euglossine bees in a forest fragment, plants that depend upon them for pollination will not reproduce. Likewise, plants that depend upon

Figure 15-1 Rainforest edge along deforested land.

monkeys or frugivorous understory birds for dispersing seeds will no longer have this service available if those animals are not entrapped within a fragment. Any offspring produced by such plants would result in seeds sprouting from fruits dropped by the parent plant with whom the offspring would be forced to grow in competition. Trees are very long lived, however, and it could be years if not decades before noticeable decline of the fragments occurred.[5]

Plants capable of self-pollination within a forest fragment would be at an advantage over those limited to cross-pollination. Cross-pollinating plants would be able to reproduce if a pollinator species was present in adequate numbers to maintain itself, but both types would face similar problems in dealing with fractions of the gene pools of which they were once part. Over time, inbreeding could become a problem. Again, given the longevity of plants, it is more likely that the fragment would be destroyed or would become eventually reunited with standing forest before anything noticeable would occur, but research has shown that tropical dry forest trees grown in isolation have less genetic variability than those in the intact forest.[6] Among animals, reproducing populations of smaller ones such as insects, other arthropods, and small vertebrates, would be more likely to be trapped within a fragment and would probably persist longer than larger species. Understory birds, for example, show population declines of 50% in 50 years in 1000 hectare (roughly 2500 acre) fragments.[7] One would anticipate that population declines of all animals would occur more quickly in smaller fragments.

One further thought on that note is that if the fragment and its relict populations were able to persist long enough, many of the organisms within it would be reproductively isolated from their conspecifics in other fragments and in the continuous forest, a situation analogous to organisms confined to islands surround by seas. Thus there would also be fragments of larger gene pools facing different selective pressures in the fragments than in the continuous forest, which would encourage genetic diversification and perhaps rapid speciation. This would be a fascinating thing to study from the perspective of evolutionary biology, but it probably wouldn't be worth the loss of species, including potentially useful ones, that fragmentation would cause.

All may not be totally bleak, however. There are conservation steps that can be taken to reduce and perhaps in some cases even prevent species loss in forest fragments. These are **wildlife corridors**, strips of undisturbed forest stretching between fragments, or preferably between fragments and continuous forest, that allow for the movement of animals, including pollinators and seed dispersers. Studies in Australia, for example, have shown that the diversity of arboreal mammals is higher in forest fragments that are linked to continuous forest by way of corridors, and the wider and more diverse the corridor, the greater the variety of mammals within the fragment.[8] Even more promising in terms of conservation of some species may be the maintenance of wildlife corridors along rivers. Scientist in Brazil found that species of small mammals and litter frogs in linear corridors along rivers were as essentially identical to those in continuous forest.[9] Whether that diversity can be maintained over time is yet to be determined, but at this point it at least provides some hope.

The reasons behind deforestation have been largely economic. Removal of commercially valuable trees and converting forest to farmland has been seen as profitable. Studies have shown that sustainable forestry can be competitive with and even more profitable than farming under the right conditions, but the nature of sustainable forestry is such that it provides returns only periodically, and many landowners prefer to have regular annual returns even if they're smaller in total.[10] In terms of preserving continuous forest, there are drawbacks to sustainable forestry. It means that mature trees will be harvested periodically and roads will be cut into the forests, essentially causing artificial gaps and streaks in the forest. In other words, some kind of disturbance is unavoidable. However, a managed, periodically harvested standing tropical forest is preferable to a barren field where forest once stood, economically and aesthetically. Moreover, it is as unrealistic to expect landowners to abandon all potential income from their property as it is to demand that governments take possession of all remaining standing forests and prohibit exploitation. The former is unreasonable and the latter would be unenforceable. If large tracts of continuous tropical forest are to be conserved, reasonable compromises have to be made.

Reforestation

A number of years ago, more than I care to count actually, I found myself in an argument with a retired farmer. He had spent his life growing crops and raising animals for human consumption, and his attitude was that there was no other reasonable use for land. When it came to forests, he believed the native trees should be cut down and replanted with timber species. Moreover, he totally rejected my points about **phytochemicals** or other forest compounds having any medicinal or other useful values, arguing instead that any useful chemical could and would eventually be synthesized in laboratories. I understand that a few years later, his wife developed cancer and was successfully treated with Taxol, a chemotherapeutic agent originally isolated from the Pacific yew, a tree found in the western United States. I sometimes wonder if that changed his mind.

The thoughts that man expressed were not that unusual. For many years it was a common practice in North America and Europe to plant commercially important trees in forests that had been overly logged. There are still forests one can walk through that are essentially single species plantations, or **monocultures,** with the dearth of biodiversity that one would expect to find in such environments. The practice has been extended to the tropics. For example, cypress and Caribbean pine have been planted in degraded forests in Uganda for wood production. The practice is apparently now being abandoned, as the exotic conifers are being harvested, and natural forests are being allowed to regenerate. Study has shown that biodiversity is much greater in the natural forests than in the plantations.[11]

Still, it is unrealistic to think that plantations are not going to be part of the future of the tropical forests. As petroleum gets more and more expensive, for example, there will be more and more pressure for biofuels, and so plantations of oil-yielding palms can be expected to become commonplace in the tropics. Furthermore, one can expect demand for timber and food crops to increase. Admittedly, the promise of a financial return may very well motivate some landowners to reforest their logged-over or overgrazed or depleted property. Tree monocultures are not ideal replacements for cleared tropical forest, but they are preferable to land lying bare and being eroded by wind and rain. Consequently, reforestation projects resulting in plantations of trees yielding commercial products can be expected to increase, and sound forest management is going to be necessary to maximize yield while trying to maintain environmental quality and biodiversity. Toward that end, there are some definitive steps that can be taken.

Work in Indonesia has demonstrated that forests that have been cleared of timber trees and are seemingly recovering through ecological succession can be encouraged to grow commercially valuable trees by creating artificial gaps that are then planted with seedlings of the desired tree species before successional species can reestablish themselves. By controlling the size of the gap, light conditions can be manipulated to the needs of the planted species.[12] It is important, however, that proper evaluation of the environmental conditions are carried out, and the trees planted are suitable for the soil in which they are being planted.[13] Planting a grove of oil palms in soil that won't adequately support them accomplishes nothing. Furthermore, native trees should be considered in reforestation projects, as these are the most likely to support other native organisms and, consequently provide for enhancing biodiversity.[14]

In Closing

In 1988, University of Pennsylvania biologist D. H. Jansen wrote, "Tropical conservation biology is inescapably the biology of habitat fragments."[15] He went on to describe those fragments (Figure 15-2) as "decomposing habitat

Figure 15-2 Forest clearing has not only destroyed much of the rainforests, it has also changed them greatly increasing the relative amounts of forest edge to intact forest.

fragments," remnants and relics of larger tracts of continuous forest that have not yet been cut down, and "growing habitat fragments," or newly reclaimed and reforested areas that environmental managers of some kind are trying to maintain either for commercial or other reasons. In an earlier essay he describes some organisms in the remnant habitats as "living dead." They appear to be very much alive and functional, but they are doomed because of their inability to reproduce, as described earlier in this chapter. The symbionts upon whom they depend for reproduction are unable to cross the barriers between continuous forest and the fragments.[16] Therefore, in order to reestablish tropical forests, conservationists will have to provide not only space and seedlings, but access to continuous forest as well, preferably by wide corridors.

When one looks at the current state of the tropical forests, and the rest of the planet for that matter, it's easy to become disheartened. Governments ignore festering problems or contribute toward making them worse. Multinational corporations likewise appear to be contributing to the environmental issues that scientists continually warn us we will have to face during this century. It's easy to envision the planet 50 years from now with the tropical forests destroyed, the coral reefs bleached, the polar ice caps melted, and coastal cities inundated by sea water. I wish I could be reassuring and say for certain that this will not happen, but I can say that there are forces at work trying to prevent it. On the governmental scale, unique land is being set aside for parks and preserves, as for example the Kaa-lya del Gran Chaco National Park in Bolivia.[17] This is an integrated park that will involve community management and allow use, with the effort being aimed at conservation rather than preservation. Efforts are being made to involve indigenous peoples in projects that make it in their best interest to conserve their tropical forests. Many countries do pass laws aimed at conservation, although enforcement has been and continues to be difficult. But grass roots efforts exist as well, sometimes coming from odd sources. I once participated in planting cecropia trees in Costa Rica on the property of a lodge at which I was staying. The owners of the lodge took it upon themselves to start reforesting their property. In Central America, Catholic clergy have been advocating the opposition of their parishioners to multinational mining companies that want to displace villages and extract gold. These actions have not been totally appreciated in Rome.[18]

Sometimes change comes about because there's no choice. When the Soviet Union collapsed and was no longer able to provide Cuba with petroleum and agricultural chemicals, Cubans learned how to deal with less fuel and without agrochemicals. They began riding bicycles and cultivating small, organic urban gardens. As a result, they ended up producing more and better quality food, as well as reducing their fossil fuel consumption and carbon dioxide generation.[19] The destruction of the tropical forests has come about at least in part from the availability of cheap energy to run chain saws and farm machinery and petrochemicals for producing insecticides and fertilizers. Agriculture through the tropical world has grown as the forests have shrunk. The demand for timber has also fueled the logging of tropical forests as well as opening them for settlement and agriculture. We now see energy getting more and more expensive. Whether it will have the same effect on other tropical countries that it had on Cuba remains to be seen. I personally doubt it, but I could be wrong. In fact, I'd like to be wrong about that.

One final thought: the eruptive growth of the world's human population over the past century has also contributed to tropical forest destruction as people in tropical overcrowded countries have destroyed forests to gather fuel wood, grow food, and simply find a place to live. Worldwide fertility control and family planning would help take pressure off the tropical forests. In all, there is much that can be done to try to preserve and restore these vital ecosystems. The question that remains is, will enough be done? Only time will answer that.

References

1. Steadman, D. W., P. S. Martin, R. D. E. MacPhee, A. J. T. Jull, H. G. McDonald, C. A. Woods, M. Iturralde-Vincent, and G. W. L. Hodgins. 2005. Asynchronous extinction of late Quaternary sloths on continents and islands. *Proceedings of the National Academy of Sciences of the United States of America* 102(33):11761–11765. http://www.pnas.org/cgi/reprint/102/ 33/11763?maxtoshow=&HITS=10&hits= 10&RESULTFORMAT=&fulltext=%28David+AND+Steadman%29&searchid=1& FIRSTINDEX=0&resourcetype=HWCIT.

2. Huxtable, R. J. 1992. The pharmacology of extinction. *Journal of Ethnopharmacology* 37:1–11.

3. Bierregaard, R. O., Jr., T. E. Lovejoy, V, Kapos, A. A. dos Santos, and R. W. Hutchings 1992. The biological dynamics of tropical rainforest fragments: A prospective comparison of fragments and continuous forest. *BioScience* 42(11):859–866.

4. Laurance, W. F., L. V. Ferreira, J. M. Rankin-de Merona, and S .G. Laurance. 1998. Rain forest fragmentation and the dynamics of Amazonian tree communities. *Ecology* 79(6):2032–2040.

5. Benitez-Malvido, J. 1998. Impact of forest fragmentation on seedling abundance in a tropical rain forest. *Conservation Biology* 12(2):380–389.

6. Cascante, A., M. Quesada, J. J. Lobo, and E. A. Fuchs. 2002. Effects of dry tropical forest fragmentation on the reproductive success and genetic structure of the tree *Samanea saman*. *Conservation Biology* 16(1):137–147.

7. Brooks, T. M., S. L. Pimm, and J. O. Oyugi. 1999. Time lag between deforestation and bird extinction in tropical forest fragments. *Conservation Biology* 13(5):1140–1150.

8. Laurance, S. G., and W. F. Laurance. 1999. Tropical wildlife corridors: Use of linear rainforest remnants by arboreal mammals. *Biological Conservation* 91(2-3): 231–239.

9. De Lima, M. G., and C. Gascon. 1999. The conservation value of linear forest remnants in central Amazonia. *Biological Conservation* 91(2-3):241–247.

10. Howard, A. F., and J. Valerio. 1996. Financial returns from sustainable forest management and selected agricultural land-use options in Costa Rica. *Forest Ecology and Management* 81:35–49.

11. Fimbel, R. A., and C. C. Fimbel. 1996. The role of exotic conifer plantations in rehabilitating degraded tropical forest lands: A case study from the Kibale Forest in Uganda. *Forest Ecology and Management* 81:215–226.

12. Tuomela, K., J. Kuusipalo, L. Vesa, K. Nuryanto, A. P. S. Sagala, and G. Adjers. 1996. Growth of diptocarp seedlings in artificial gaps: An experiment in logged-over rainforest in South Kalimantan, Indonesia. *Forest Ecology and Management* 81:95–100.

13. Butterfield, R. P. 1996. Early species selection for tropical reforestation: A consideration of stability. *Forest Ecology and Management* 81:161–168.

14. Butterfield, R. P. 1996. 1995. Promoting biodiversity: Advances in evaluating native species for reforestation. *Forest Ecology and Management* 75:111–121.

15. Janzen, D. H. 1988. Management of habitat fragments in a tropical dry forest: Growth. *Annals of the Missouri Botanical Garden* 75(1):105–116.
16. Janzen, D. H. 1986. The future of tropical ecology. *Annual Review of Ecology and Systematics* 17:305–324.
17. Taber, A., G. Navarro, and M. A. Arribas. 1997. A new park in the Bolivian Gran Chaco—An advance in tropical dry forest conservation and community-based management. *Oryx* 31(3):189-198.
18. Snell, M. B. 2007. Bulldozers & blasphemy. *Sierra* October:36–43,63,76.
19. Mark, J. 2007. Growing it alone. *Earth Island Journal* 22(1):32–36.

Glossary

Abiotic: Literally without life. When describing resources, this generally refers to non-living resources such as mineral ores, water, and fossil fuels. In ecological terms, it refers to non-living factors in an organism's environment, such as soil fertility, rainfall, and average annual temperature.

Alluvial: Deposits left by moving water.

Adventitious roots: Roots that originate in places on plants other than at the very base of the stem.

Anaerobic: Without air. This usually refers to metabolic processes carried on by organisms in the absence of oxygen.

Angiosperm: A flowering plant.

Anthropogenic: Caused by people.

Anoxic: Devoid of oxygen.

Aquifers: Underground layers of porous rock that hold water.

Arboreal: Tree-dwelling.

Batesian mimicry: Named for British scientist Walter Bates, this describes a nontoxic or otherwise nondangerous animal resembling one that is toxic or otherwise dangerous. A common example is the nontoxic viceroy butterfly of North America resembling the toxic monarch butterfly.

Biodiversity: The variety of organisms found in a defined area or region.

Biofuel: A source of energy extracted from living organisms. An example would be ethanol made from corn.

Biome: A biological region that is defined by particular climatic or physical factors.

Biotic: Literally living. When applied to resources, this generally refers to living resources such as trees, fish stocks, and harvestable foods. In ecological terms, it refers to living factors in an organism's environment, such as predators, competitors, and food organisms.

Blackwater river: A river that carries little or no sediment. In the tropics these are often dark due to dissolved humic material.

Canopy: The uppermost layer of a forest; the level that includes the high branches and crowns of trees.

Chiropterophilous: Literally bat-loving, refers to plants that are pollinated by bats.

Chlorophyll: The green pigment in plants that is essential for photosynthesis to occur.

Clear-cut: In reference to a forest, the cutting down of all trees.

Climax forest: See **old growth forest.**

Coevolution: An interactive influence two organisms have on the change of genetic composition of one another. In general, it results in mutual dependence of each on the other.

Commensalism: A relationship between two organisms where one benefits and the other is unaffected.

Community: In the ecological context, the collection of organisms found within a defined habitat.

Conspecific: Members of the same species.

Convergent evolution: The evolution of similarity among disparate organisms that occupy similar niches in different ecosystems.

Coriolis effect: In the context of this book, the deflection of sinking air particles by the spin of the Earth upon striking the Earth.

Deciduous: In biology, used to describe structures that fall from the organism that grew them. In botany, it is used to describe trees that drop their leaves during the dormant season.

Decomposition: The decay of organic material.

Definitive host: In parasitology, the host in which an internal parasite completes sexual reproduction.

Divergent evolution: See **speciation**

Ecological equivalent: Organisms occupying identical niches in different habitats.

Ecological succession: The process of change in species of organisms following disturbance to a habiat.

Ecotourism, also know as ecological tourism: A form of tourism that appeals to ecologically and socially conscious individuals. It conserves the envionment and improves the well-being of local people.

Ecosystem: In the ecological context, the system resulting from the collection of organisms and nonliving factors found within a defined habitat.

Ecotone: A transition zone between two ecosystems. It often contains organisms common to both ecosystems plus some that are unique to itself.

Ectoparasite: A parasite that lives on the outer surface of its host.

El Niño: Also known as the Southern Oscillation, this is an accumulation of warm water off the west coast of South America that affects wind, rainfall, and temperature over much of the globe. The term is Spanish for "the child."

Emergents: In the context of the tropical rainforests, tall trees whose crowns stand above the canopy.

Endoparasite: A parasite that lives within its host.

Epiphyte: A plant that is physically supported by another plant from which it receives no nutrition.

Filter route: A migration pathway that allows some organisms to cross but inhibits others.

Forest floor: The ground surface in a forest.

Fragmentation: Whenever an ecosystem is disrupted in such a way that isolated parts of it remain undisturbed.

Frugivory: Fruit-eating.

Gondwanaland: The supercontinent composed of Africa, Antarctica, Australia, India, and South America that existed approximately 500 to 550 million years in the past.

Greenhouse effect: The phenomenon behind global warming.

Groundwater: Underground reservoirs of water.

Guttation: The loss of water in liquid form from leaves.

Gymnosperm: Nonflowering seed plants.

Halophile: Literally a salt-loving plant, it is a plant that grows in saline conditions.

Herbivore: A plant-eating animal.

Horizon: In soil science, a layer of soil.

Hydrologic cycle: The route followed by water in circulating between the living and nonliving worlds.

Intertropical Convergence Zone: A permanent zone of low pressure encircling the Earth at the equator; it is the convergence of the northeast and southeast trade winds.

Jungle: Often used synonymously with tropical rainforest, it is seemingly tangled, impenetrable tropical forest, usually found at the forest edge or in areas of succession.

Keystone species: A particularly important species whose presence in an ecosystem influences the biological structure of the ecosystem.

Leaching: The loss of nutrients from soil by water passing through it.

Liana: A woody vine.

Limiting factor: A factor in an ecosystem that limits the population growth of species within it.

Litter: In soil science, the layer of leaves and other organic debris on top of the soil.

Mangal: The collection of trees and plants in a mangrove swamp or forest.

Mantel: In earth science, the molten layer of rock found between the Earth's crust and the core.

Melanism: A characteristic of many kinds of vertebrates in which a genetic variation causes an individual to be black rather than the typical color.

Monoculture: The growth of a single cash crop.

Monsoon climate: A tropical climate characterized by a heavy rainy season alternating with a very dry season.

Mullerian mimicry: Named for German naturalist Fritz Müller, this describes two or more toxic or otherwise dangerous animals resembling one another. The relationship reinforces the toxic warning of the species to potential predators

Mutualism: A relationship where two species of organisms live together for mutual benefit.

Mutualistic symbiosis: See **mutualism**

Mycorrhizea: Soil fungi that are symbiotic with the roots of trees. These organisms absorb water and mineral nutrients from the soil and pass it on to the trees. In turn, the trees provide energy nutrients.

Myrmecophyte: Literally, an ant plant. In the context of this book, a symbiotic relationship between a plant and a population of ants that it houses. In general, the plant provides living space and secretes nutrients while the ants provide protection.

Nich: The role that an organism happens to play and how it interacts with other organisms in its community or environment.

Old growth forest: A forest that has remained undisturbed for many years and is characterized by large, mature trees.

Parasitism: A symbiotic relationship between two organisms where one benefits by harming the other.

Pheremone: A chemical secreted by one animal that stimulates a behavioral response in another.

Phytochemical: A chemical isolated from a plant.

Photosynthesis: A light-driven chemical process of green plants where chlorophyll synthesizes glucose from carbon dioxide and water.

Pioneer community: The first assemblage of plants to grow following the destruction of a habitat.

Plate tectonics: The theory that explains the movement of continents across the Earth's surface.

Pneumatophore: In some plants with submerged roots, a vertical outgrowth of the root that extends above the water's surface and allows for gas exchange between the root and atmosphere.

Preservation: When no natural resources are drawn from an area.

Primary forest: See **old growth forest.**

Primary succession: The return of an ecological community to a habitat following complete destruction.

Prop root: An adventitious root growing off the lower trunk of a tree that helps support the tree.

Prosimian: Refers to primates that are considered to be ancestral to and more primitive than the monkeys and apes.

Pteridophyte: Vascular plants that reproduce by spores rather than by flower and seed. In particular, the ferns.

Quinine: A medication to fight malaria first isolated from the bark of tropical New World cinchona trees.

Rain Shadow: A dry region usually downwind from a geological formation such as a mountain that blocks moisture.

Respiration: A series of energy-liberating chemical reactions carried on by organisms. In most cases it consumes oxygen and releases carbon dioxide.

Rhizome: An underground stem.

Saline: As an adjective, salty, as in saline water. As a noun, a 0.9% solution of salt in water.

Secondary forest: A forest undergoing succession, i.e., one that has not yet reached old growth stage.

Secondary succession: Succession beginning at a point later than the pioneer stage, usually when a biological community has been disturbed but not destroyed.

Sexual mimicry: Where the characteristics of one sex in a species of animals resembles the other.

Shaman: A medicine man.

Silverback: Dominant adult male in a group of gorillas.

Speciation: The evolutionary change resulting from the isolation of one population of a species from another. Eventually, the isolated population is no longer recognized as the same species.

Succession: The sequential change in a biological community following disturbance or destruction.

Successional forests: Forests recovering from some kind of disturbance and progressing toward becoming old growth.

Sustainability: Agricultural practice that removes plant products at a rate no faster than they are naturally replaced. This is sometimes referred to as **sustainable agriculture.**

Sustainable travel: Travel that does not impact an environment beyond its capacity to repair itself.

Symbiosis: Literally living together, a relationship between two organisms.

Transpiration: The evaporation of water from leaves.

Tropics: Any portion of the earth characterized by a tropical climate.

Understory: In a forest, the layer of plants of intermediate height; those that do not grow tall enough to reach the canopy.

Weathering: The breakdown of rock to smaller particles, a process in soil formation.

Whitewater river: In this context, sediment-laden, usually nutrient-rich rivers.

Wildlife corridors: Strips of undisturbed forest stretching between fragments, or preferably between fragments and continuous forest, that allow for the movement of animals, including pollinators and seed dispersers.

Xeric: Dry conditions.

Index

NOTE: The italicized *f* following a page
number indicates a picture will be found on that page.